# Ferrari 250/GT

Service and Maintenance for the
250GT, 250GTE, 250GTL, 250SWB
and other similar models

*Complied by:* JIM RIFF

USO E MANUTENZIONE

# INDEX

| | |
|---|---:|
| FOREWORD | 3 |
| INTRODUCTION | 4 |
| GENERAL SPECIFICATIONS | 5 - 12 |
| ELECTRICAL | 13 - 38 |
| CARBURETTORS | 39 - 54 |
| ENGINE | 55 - 78 |
|     CYLINDER HEADS | 56 - 59 |
|     CRANKSHAFT | 60 - 62 |
|     CAMSHAFTS | 63 - 67 |
|     PISTONS AND RINGS | 68 - 69 |
|     VALVE ADJUSTMENT | 70 - 72 |
|     LUBRICATION SYSTEM | 73 - 74 |
|     COOLING SYSTEM | 75 - 78 |
| GEARBOX AND CLUTCH | 79 - 88 |
| OVERDRIVE | 89 - 108 |
| BRAKES | 109 - 114 |
| REAR AXLE, DRIVE SHAFTS, SHOCK ABSORBERS AND STEERING | 115 - 122 |
| ALIGNMENT, WHEELS AND TIRES | 123 - 130 |
| TORQUE SPECIFICATIONS AND METRIC TO USA CONVERSION DATA | 131 - 134 |
| PAINT CODES | 135 - 137 |
| EPILOG by JIM RIFF | 138 |

# **FOREWORD**

One of the fringe benefits of being a publisher is that you get to meet many interesting individuals. While they all seem to have wonderful anecdotes of the celebrities that they have met or worked with, it is their enthusiasm and technical knowledge for the subject that invariably gets my attention.

In a recent phone conversation with Richard Merritt, Dick and I were discussing the reprinting of his book "Ferrari: Operating, Maintenance and Service Handbooks". My normal process during these phone conversations is to listen and scribble notes for review later, and one of my notes, heavily circled, was "250GT manual rare" a comment Dick had made while discussing his book. After the phone conversation was over, I pulled a copy of Dick's book from the shelf to review page layout, margins etc., the necessary but mundane side of the publishing business. However, as is normal with most books, it seems I always end up reading them and this was no exception. I was only a few pages into Dick's book when I came upon the list of acknowledgments and contributors and one name jumped from the page, that name was Jim Riff. It seems that almost every book that deals with the technical side of 60's and 70's Ferrari's either gives credit or names Jim as a contributor. Including Dick Merritt's "Ferrari: Operating, Maintenance and Service Handbooks", Kurt Miska's "Berlineta Lusso", and of course Jim's original version of "Ferrari Tuning Tips and Maintenance Techniques" which was later updated by John Apen and Gerald Roush. The reason why Jim's name made such an impact on me was that I remembered we had discussed his "250GT Service and Repair Manual" a few years previous but for some reason had never got around to publishing it, and there right in front of me was my note "250GT manual rare" so I made another note to contact Jim after we had finished reprinting Dick's FOMSH.

Shifting gears to a few weeks after my phone conversation with Dick, I received an unsolicited email from Jim Riff asking if we had any interest in publishing his 250GT Service Manual, it seemed that fate and coincidence were hard at work! Obviously, I agreed and asked Jim to send me the originals so that we could get the book into pre-press as quickly as possible. A few days later I received a package from Jim and eagerly tore it open, my reaction must have been interesting to witness, as the packaged contained approximately 90 pages of typed (remember typewriters?) text that included many illustrations, diagrams and drawings. However, the worse part was that there were no originals! These were photocopies of photocopies and anyone who has seen a document that has been copied over and over again will no doubt understand my grief, how was I ever going to be able to resurrect these documents to a point that they would be good enough to publish as a book? Important data had slipped off the edges, illustrations were skewed, and many of the notes to the illustrations were hand-written by Jim and no longer decipherable, but the technical data and information contained in these pages was nothing short of incredible.

My initial reaction was just to have someone re-key the "originals" but the more I looked through the pages the more I realized what an impossible task it would be. Also, there was a certain "feel" that would be destroyed if the whole thing were digitized. Therefore, I decided to work with what Jim had sent us and retain as much of the look and feel of his original effort as possible. What you now hold in your hand is the result of that decision and while some readers may be critical of the presentation, it is a tribute to what can be done with digital technology and should anyone doubt that statement please come visit with us to see what we started with!

Dave McClure ~ Publisher ~ www.VelocePress.com

## 250 SERIES WORKSHOP MANUAL

A collection of technical materials intended to assist owners in the care and repair of the 250 series of Ferrari vehicles. Chapters include engine, chassis, electrical, alignment, fuel systems, running gear, paint codes, cooling systems, brakes, ignition and torque specifications.

The works are intended as a guide and reference for these vehicles and to assist owners as to the details of a specific repair in order to help in making informative decisions on the task. Covers 250GT, 250GTE, 250GTL, 250SWB and other similar models. Over 100 pages of applicable materials and notes.

*Jim Riff ~ 2006*

## NOTE FROM THE PUBLISHER

The information presented is unchanged from the original compilation and has not been updated to reflect changes in common practice, new technology, availability of improved materials or increased awareness of chemical toxicity. As such, it is advised that the user consult with an experienced professional prior to undertaking any procedure described herein.

## INFORMATION ON THE USE OF THIS PUBLICATION

This compilation is an invaluable resource for both the Ferrari enthusiast and owner interested in performing their own maintenance. The reader will find repair and service data and step-by-step instructions and illustrations on dismantling, overhauling, and re-assembly. Certain assemblies require the use of expensive special tools and although repair information is included, it is recommended that these repairs be performed by factory authorized service centers.

Whilst every care has been taken to ensure correctness of information, it is obviously not possible to guarantee complete freedom from errors or omissions or to accept liability arising from such errors or omissions. Therefore, by using the information contained within this manual, any individual that elects to perform or participate in do-it-yourself repairs acknowledges that there is a risk factor involved and that the publishers or its associates cannot be held responsible for personal injury or property damage resulting from the outcome of such repairs.

It is important that the reader recognizes that these instructions may refer to either the right-hand or left-hand sides of the vehicle or the components and that the directions are followed carefully. One final word of advice, this compilation is intended to be used as a reference guide, and when in doubt the reader should consult with a qualified Ferrari expert.

## 250GT ~ GENERAL SPECIFICATIONS

| | |
|---|---|
| PRINCIPAL CHARACTERISTICS | 6 |
| TUNE UP SPECIFICATIONS | 7 |
| ROUTINE MAINTENANCE | 8 |
| CAPACITIES | 9 |
| RECOMMENDED LUBRICANTS | 10 - 11 |
| IGNITION SYSTEM - DESCRIPTION | 12 |
| LUBRICATION SYSTEM - DESCRIPTION | 12 |
| COOLING SYSTEM - DESCRIPTION | 12 |

## PRINCIPAL CHARACTERISTICS

### MOTOR

| | |
|---|---|
| Number of Cylinders | 12 |
| Arrangement of Cylinders | Vee 60° |
| Bore and Stroke | 73mm X 58.8 mm |
| Displacement | 2953 cmc/180 ci. |
| Compression Ratio | 9.2 |
| Horsepower at 7,000 RPM | 235 din |

### GENERAL

| | |
|---|---|
| Wheel Base | 2,600 mm/ 8.53 ft. |
| Maximum Length | 4,700 mm/ 15.42 ft. |
| Maximum Width | 1,710 mm/ 5.61 ft. |
| Curb Weight | 1,310 kg/ 2,889 lbs. |
| Total Weight (4 Persons) | 1,695 kg/ 3,737 |
| Maximum Speed | 230 kph/ 143 mph |
| Body Style | Coupe |
| Number of Passengers | 2 + 2 |
| Electrical System | 12 Volt |
| Generator | 30 amperes |
| Wheels | RW3690 (6.00 X 15") |
| Tires | Pirelli  185/ 15 |

### REFERENCE

| | |
|---|---|
| Fuel Capacity | 90 Li/ 23.8 Us. gals. |
| Reserve Fuel Capacity | 16 Li/ 4 Us gal. |
| Radiator Capacity | 11.5 Li/ 3.1 Us gal. |
| Engine Oil | 10 Li/ 10.6 qts. |
| Gearbox and Od. | 4.4 Li/ 9.3 pints |
| Rear End | 1.8 Li/ 3.8 pints |
| Steering Box | .4 Li/ .85 pints |
| Brake Fluid | .81 Li/ 1.7 pints |
| Front Shocks | .32 Li/ .68 pints |
| Rear Shocks | .35 Li/ .74 pints |

## 250GT TUNE UP SPECIFICATIONS

### ELECTRICAL

| | |
|---|---|
| Ignition Point Gap | .014" |
| Ignition Capacitor | .18ufd |
| Spark Plug Type | N3 Champion |
| Spark Plug Gap | .020" |
| Timing | 10° BTDC at 800 RPM |
| Voltage Regulator Setting | 14.20 ± .20 volts |

### VALVE CLEARANCE (Lash)

| | |
|---|---|
| Intake | .006" cold |
| Exhaust | .008" cold |

### CARBURETOR

| | |
|---|---|
| Idle Speed | 600 - 800 RPM |
| Fuel Pressure | 3 PSI (min) |
| Float Level | 3 MM |
| Main Jet | 1.40 MM |
| Slow Run | .60 MM |
| Pump | .60 MM |
| Air Correctors | 2.40 MM |
| Chokes | 27 MM |
| Fuel Pump (Engine) | Fispa Sup. 150 |
| Fuel Pump (Electric) | Fispa PBE 10 |

### WHEEL ALIGNMENT

| | |
|---|---|
| Toe Out | .06" |
| Camber | 1/4" to 3/8" (1°) |
| Tire Pressure | 28 PSI front, 33 PSI rear |
| Front Brake Pads | MINTEX VBO - 875/5201 |
| Rear Brake Pads | MINTEX VBO - 875/5138 |

## ROUTINE MAINTENANCE

| | | |
|---|---|---|
| Before using the car: | 1. | Check water level in the radiator. |
| | 2. | Check oil level in the sump. |
| | 3. | Check tire pressure. |
| | 4. | Check brake fluid level. |
| Every 300 miles: | 5. | Check or refill radiator fluid level. |
| | 6. | Adjust tire pressure. |
| Every 1,500 miles: | 7. | Check electrolyte in battery. |
| Every 3,000 miles: | 8. | Check tension of fan belt. |
| | 9. | Clean carburetor air filter. |
| | 10. | Check brake pads and pedal movement. |
| | 11. | Rotate tires. |
| | 12. | Clean and adjust breaker points. |
| Every 6,000 miles: | 13. | Replace spark plugs. |
| | 14. | Check valve clearance. |
| | 15. | Adjust timing chain tension. |
| | 16. | Replace brake pads and bleed. |
| | 17. | Check action of shock absorbers. |
| | 18. | Check starter motor brushes and commutator. |
| | 19. | Adjust clutch pedal movement. |
| | 20. | Take up steering play. |
| Every 12,000 miles: | 21. | Check wheel toe-in and camber. |
| | 22. | Clean fuel filters. |
| | 23. | Check carburetors and controls. |

Note:

Front wheel toe-in and camber must be checked whenever the car is involved in a collision. In the event of damage, the steering links should be replaced as their reconditioning is not possible.

## CAPACITIES

**Replenishing**

|  | Litres | Gallons/Pints |  |
|---|---|---|---|
| **WATER** | 11 | 2 | 3 |
| **FUEL** | 90 | 19 | 7 |
| **OIL** | Kilograms. | | |
| Sump | 9 | — | 15 |
| Filters | 1 | — | 2 |
| Gearbox & Overdrive | 4.6 | — | 6 |
| Rear axle | 1.8 | — | 5 |
| Steering box | 0.4 | — | 1 |

| LUBRICANTS ||
|---|---|
| Cambio<br>Boite de vitesse<br>Gear Box | Shell Spirax EP 90 |
| Overdrive | Shell Spirax EP 90 |
| Ponte posteriore<br>Pont arriere<br>Rear Axle | Shell Spirax EP 250 |
| Scatola guida<br>Boitier de direction<br>Steering box | Shell Spirax EP 140 |
| Impianto freni<br>Freins<br>Brake System | Shell Donax B SAE 70 R 3<br>Dunlop Racing Brake Fluid |
| Bracci sospensione anteriore<br>Organes de suspension avant<br>Front suspension arms<br><br>Perni fusi a snodo<br>Axes de fusees<br>Stub axle pins<br><br>Giunto cardanico trasmissione<br>Gardan de trasmission<br>Universal joint | Shell Retinax A |
| Cuscinetti a sfere ruote<br>Roulements a billes roues<br>Wheel ball bearings | Shell Retinax AX |
| Ammortizzatori<br>Ammortisseurs<br>Shock absorbers | Shell Donax A 1 |

### Filtro olio (oil filter)

L'olio di lubrificazione del motore viene filtrato dalle impurita da un filtro FRAM PH 2815 a filtraggio totale e da un secondo a filtraggio parziale PB 50.

<u>Ogni 5000 Km</u> sostituirli entrambi usando lo speciale attrezzo per svitarli dalle proprie sedi.

Accertarsi che non vi siano perdite di olio dopo la sostituzione.

15 - Lavare con petrolio le balestre posteriori.

## Lubrificanti da usare (ENGINE OIL)

| MOTORE Moteur Engine | Stagione estiva<br>Temperatura oltre i 15°C<br><br>Saison d'été<br>Temperature del plus de 15°C<br><br>Summer season<br>Temperature above 59° F | Shell X 100 SAE 40<br>Shell X 100 Multigrade 20 W 40 |
|---|---|---|
| | Stagione intermedia<br>Temperatura da - 5° a + 15°C<br><br>Saison intermédiaire<br>Temperature de -5° é + 15° C<br><br>Intermediate season<br>Temperature between 23° F to 59° F | Shell X 100 SAE 30<br>Shell X 100 Multigrade 20 W 40 |
| | Stagione invernale<br>Temperatura inferiore a -5°C<br><br>Saison invernale<br>Temperature inférieur a - 5°C<br><br>Winter season<br>Temperature lower than 23° F. | Shell X 100 SAE 20/20 W<br>Shell X 100 Multigrade 10 W 30 |

☐ Shell X 100 motor oil
◇ Shell EP 90 spirax
△ Shell Dentax 250
▽ Shell Dentax 140
▢ Shell Retinax A (Grease Fittings)
○ 20W Oil

Every 5000 miles
Every 2500 miles
Every 300 miles

## IGNITION

Firing order:   1/7/5/11/3/9/6/12/2/8/4/10.

The ignition is produced by the battery and a distributor for each line of cylinders, actuated by means of helical gears driven by the camshaft.

- automatic advance          30°
- pitch setting advance      10° + 12° (A.F.)
- max. total advance         40° + 42° (A.M.)
- contact point gap          .014"
- spark plug electrode gap   .020"

Recommended type of spark plug : Champion

## LUBRICATION

A pressure system of lubrication is provided by means of a gear pump.
A register valve maintains the proper pressure in the oil circuit.
A full-flow cartridge type oil filter.
A partial-flow, shunt mounted, cartridge type oil filter.
The normal pressure, with an oil temperature of 212-230°F at the max. engine speed of 6600 r.p.m. is 80 lb/sq.in. (mt. 60)
The minimum tolerable pressure, for the same conditions of speed and temperature is 49-56 lbs./sq. in.
The minimum pressure at lowest speed, 600 to 800 r.p.m. is 14+21 lb/sq. in.

## COOLING SYSTEM

The circulating cooling water is obtained by means of a centrifugal pump, installed on the front end of the distribution housing, and driven by the same timing chain.

The automatic control of the temperature is obtained by the thermostat located on the upper part of the radiator.
The thermostat setting is as follows:

- opening, start          80° to 84°C     176°F to 183°F
- closing, end                            194°F to 203°F

The electric fan is a three blade Marelli type SW 599 FA - 80 W - 12 V, combined with a remote control switch, Marelli type SW 599 FA/7, or a Lucas electric fan type 3GMGC-36 W - 12 V, located in the front of the radiator. (Peugeot electrically clutched engine fan).

A switch unit, installed on the lower end of the radiator, provides the automatic connection when the radiator reaches a temperature of 183°F, and disconnects automatically when the temperature drops to 167°F.

The radiator core, formed in one block, has the tubes disposed in three vertical rows and is provided with an air tight cap.

## 250GT ~ ELECTRICAL

| | |
|---|---:|
| GENERAL SPECIFICATIONS | 14 |
| RECOMMENDED SPARK PLUGS | 15 |
| WINDSCREEN WIPERS | 16 - 17 |
| IGNITION COIL | 18 |
| FUSES AND IDENTIFICATION | 19 |
| WIRING DIAGRAM - SCHEMATIC | 20 - 21 |
| WIRING DIAGRAM - COMPONENT | 22 - 23 |
| DISTRIBUTOR AND IGNITION TIMING | 24 - 32 |
| GENERATOR | 33 |
| STARTER MOTOR | 34 - 38 |

## 250GT SERIES ELECTRICAL SPECIFICATIONS

| | |
|---|---|
| System Voltage: | 12 Volts |
| Polarity: | Negative Ground |
| Battery: | 65AH 12 Volt |
| Charging Voltage: | 14.20V ± .20V |
| Generator: | Marelli DN63B-400/12/2300S |
| Generator Rating: | 30 Ampere 14 Volts |
| Voltage Regulator: | Marelli IR 19E/30012 |
| Starter Motor: | Marelli MT 21T-1.8/12D9 |
| Distributor: | Marelli S85A-12V-15° |
| Ignition Points: | Marelli H-710071-02 |
| Ignition Capacitor: | Marelli CE-1E-8E-.18MFD |
| Ignition Coil: | Marelli 12V-B202A |
| Ignition Resistor: | Marelli 1.5 Ohm |
| Windscreen Wiper: | Lucas Two Speed |
| Fuses: | 8A and 16A Lucas or Buss |
| Headlamp Assembly: | Marchal 7" No. 3-1263B |
| Headlamp Bulb: | Marchal 12 Volt No. 1263B |
| Dash Lamps | GE No. 363 |
| Electric Fuel Pump | Fimac PBE-10 |

## SPARK PLUG RECOMMENDATIONS

NOTE: Heat ranges shown are for average conditions. Torque to 22' lbs. max.

| FERRARI MODEL | PLUG GAP | CHAMPION PLUG TYPE | AC PLUG TYPE | NGK PLUG TYPE | BOSCH PLUG TYPE | AUTOLITE PLUG TYPE | KLG PLUG TYPE |
|---|---|---|---|---|---|---|---|
| 166 Norm. | .020" | N-3 | 43XL | B-7ES | W240T2 | AG2 | DFE-75 |
| 212 Inter. | .020" | N-3 | 43XL | B-7ES | W240T2 | AG2 | DFE-75 |
| 340 Amer. | .020" | N-3 | 43XL | B-7ES | W240T2 | AG2 | DFE-75 |
| 4.1 Amer. | .020" | N-3 | 43XL | B-7ES | W240T2 | AG2 | DFE-75 |
| 4.5 Comp. | .016" | N-57R | | B-8ES | W310T17 | | |
| MM Export | .016" | N-57R | | B-8ES | W310T30 | | |
| 206 Dino | .020" | | 41XLS | | | | |
| 246 Dino | .020" | N-6Y | 41XLS | B-8ES | W230T30 | AG12 | FE-125P |
| 246 GT Dino | .020" | N-6Y | 41XLS | B-8ES | W300T30 | | |
| 250 GT | .020" | N-3 | 43XL | B-7ES | W240T2 | AG2 | DFE-75 |
| 275 GTB2 | .020" | N-6Y | 43XL | B-7ES | W230T30 | AG12 | FE-125P |
| 275 GTS | .020" | N-6Y | 43XL | B-7ES | W240T30 | AG12 | FE-125P |
| 275 GTC | .020" | N-6Y | 43XL | B-7ES | W230T30 | AG12 | FE-125P |
| 275 GTB4 | .020" | N-6Y | 43XL | B-7ES | W230T30 | AG12 | FE-125P |
| 330 GT2+2 | .020" | N-6Y | 43XL | B-7ES | W230T30 | AG12 | FE-125P |
| 330 GTC | .020" | N-6Y | 43XL | B-7ES | W230T30 | AG12 | FE-125P |
| 365 GT | .020" | N-6Y | 41XLS | B-7ES | W230T30 | AG12 | FE-125P |
| 365 GTB | .020" | N-6Y | 41XLS | B-7ES | W230T30 | AG12 | FE-125P |
| 365 GTB4 | .020" | N-6Y | 41XLS | B-7ES | W230T30 | AG12 | FE-125P |
| 365 GTC | .020" | N-6Y | 41XLS | B-7ES | W230T30 | AG12 | FE-125P |
| 500 SF | .020" | N-6Y | 41XLS | B-7ES | W230T30 | AG12 | FE-125P |

# WINDSCREEN WIPERS

## 6W. WINDSCREEN WIPER.

Wiper model 6W has been developed for applications requiring greater torque than is available from model DR3. (The cold stall torque of the 6W is 1,500 ozf ins). The 6W uses the same armature and field system as the corresponding DR3. The increased power has been obtained by changing the type of gearing. The DR3 has worm and wheel gearing, while the 6W has 2 stage spur gearing. (The first stage is helical and the second stage is straight).

Both gears of the 6W wipers are moulded in Delrin — a plastic material, similar to nylon, but having increased impact strength at higher temperatures. Spur type gearing has been adopted in order to reduce transmission losses.

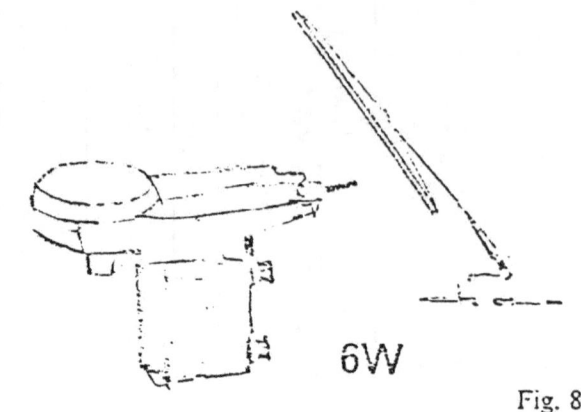

Fig. 8

## 6W. DESIGN FEATURES.

Like the DR3, the wiper has 2 speeds:—
Normal Speed: 45–50 rev/min.
High Speed: 60–70 rev/min.

In general appearance, the 6W closely resembles the DR3. However, the mounting studs are located on the yoke, and so face in the same direction as the cable rack outlet. Consequently, it is not necessary to invert the 6W wiper to fix it on either side of the vehicle.

The 6W has a "self-parking" device. When the control switch is turned to "parking", the motor rotation is reversed, and the eccentric coupling increases the movement of the wiper arm in the direction of parking, until the limit switch cuts off the supply to the motor. The wiper arms are then parked near the windscreen surround. The limit switch is built into the gearbox, and is operated by a striker on the crosshead. It is adjusted by means of a nut, near the cable outlet of the gearbox.

The 6W is used in conjunction with throated gear wheelboxes.

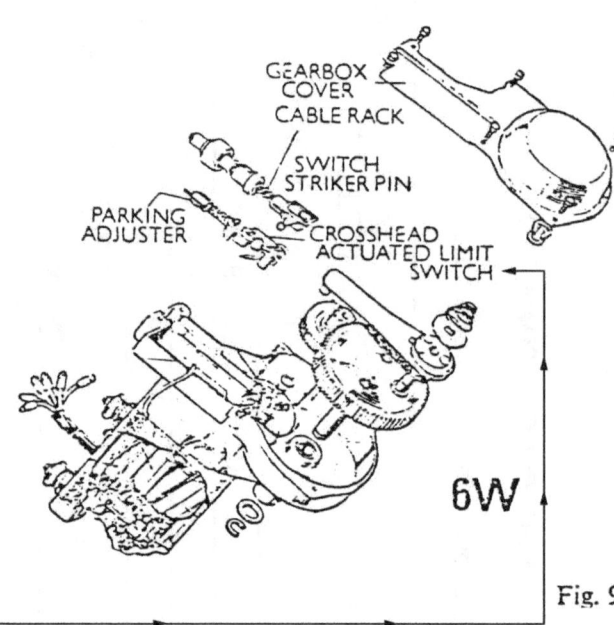

Fig. 9

## 6W. WINDSCREEN WIPER AND 79SA CONTROL SWITCH.

Figure 10 shows the internal connections of the 6W windscreen wiper, and also the 79SA control switch.

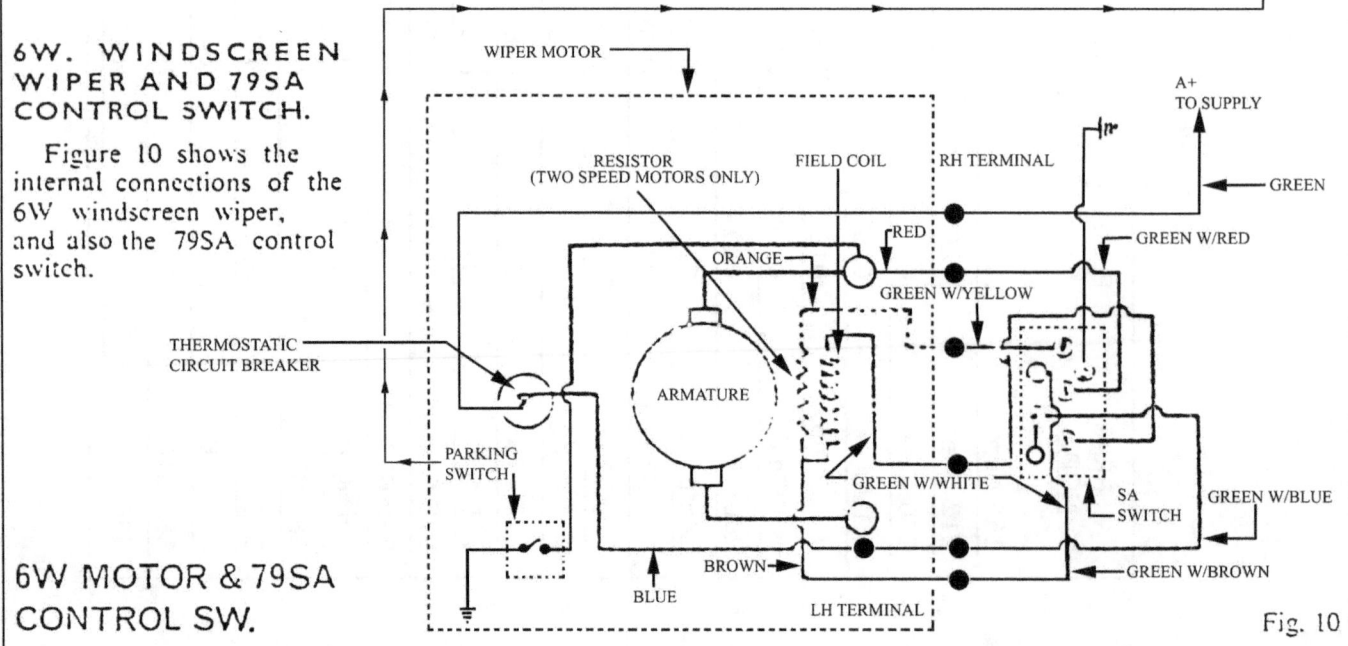

6W MOTOR & 79SA CONTROL SW.

Fig. 10

# WINDSCREEN WIPERS

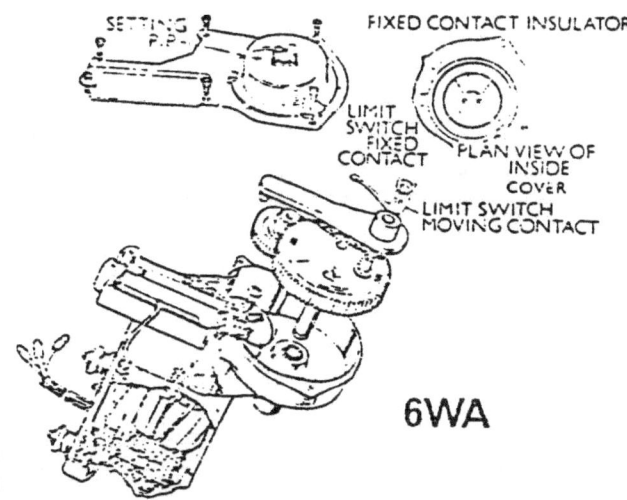

Fig. 11

## 6WA. SELF-SWITCHING.

In addition, there is the self-switching model — 6WA. This has a non-reversing motor, incorporating a limit switch to bring the blades to rest at the end of the wiping stroke.

---

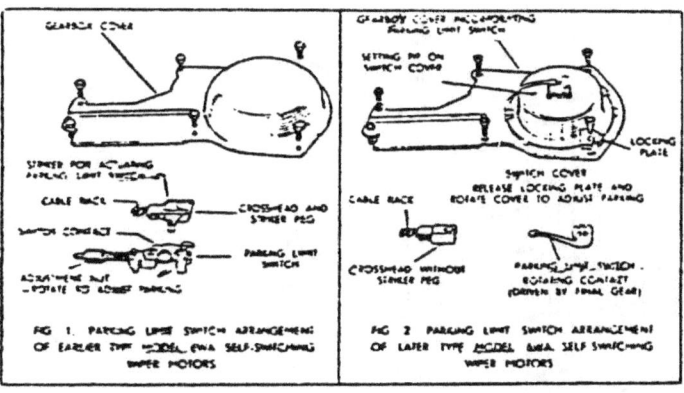

Fig. 12

## 6WA. LIMIT SWITCHES.

There are two types of limit switch used on the 6WA. The switch of earlier models was actuated by a striker peg on the crosshead of the cable rack. (Shown on left of Fig. 12).

Later model 6WA wipers have the switch housed under the domed cover of the gearbox (see Fig. 12). The cable racks for these motors are not fitted with a striker peg.

6WA motors (with rotating switch contact) can be used as a service replacement for the earlier type (with crosshead switch). The existing cable rack can still be used. (The striker on the crosshead is not required to operate the limit switch, but it can be accommodated in the space originally intended for the crosshead switch).

However, the earlier type motor cannot be used to replace the later version, unless the cable rack is also changed. Failure to do this will cause the wiper blades to park incorrectly.

---

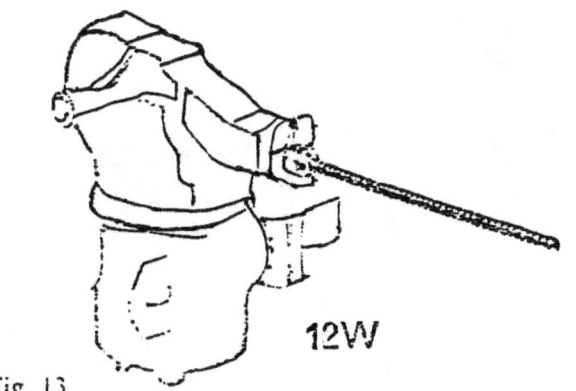

Fig. 13

## 12W. WINDSCREEN WIPER.

Instead of having wound field coils, the 12W employs a permanent magnet field system, consisting of two Barium Ferrite ceramic magnets. The 12W is more powerful than the 6W (Cold stall torque figures: 12W, 1,650 ozf ins: 6W, 1,505 ozf ins). It is also much quieter in operation, and more efficient. Consequently, the current consumption is lower for the corresponding speeds.

The 12W weighs 4·1 lbs. and is thus lighter than the 6W (5·7 lbs).

The two permanent magnets and the armature are contained in a cylindrical yoke. When the wiper is switched on, the armature rotates.

# MARELLI IGNITION COIL - PART NUMBER 12V BZR 201A

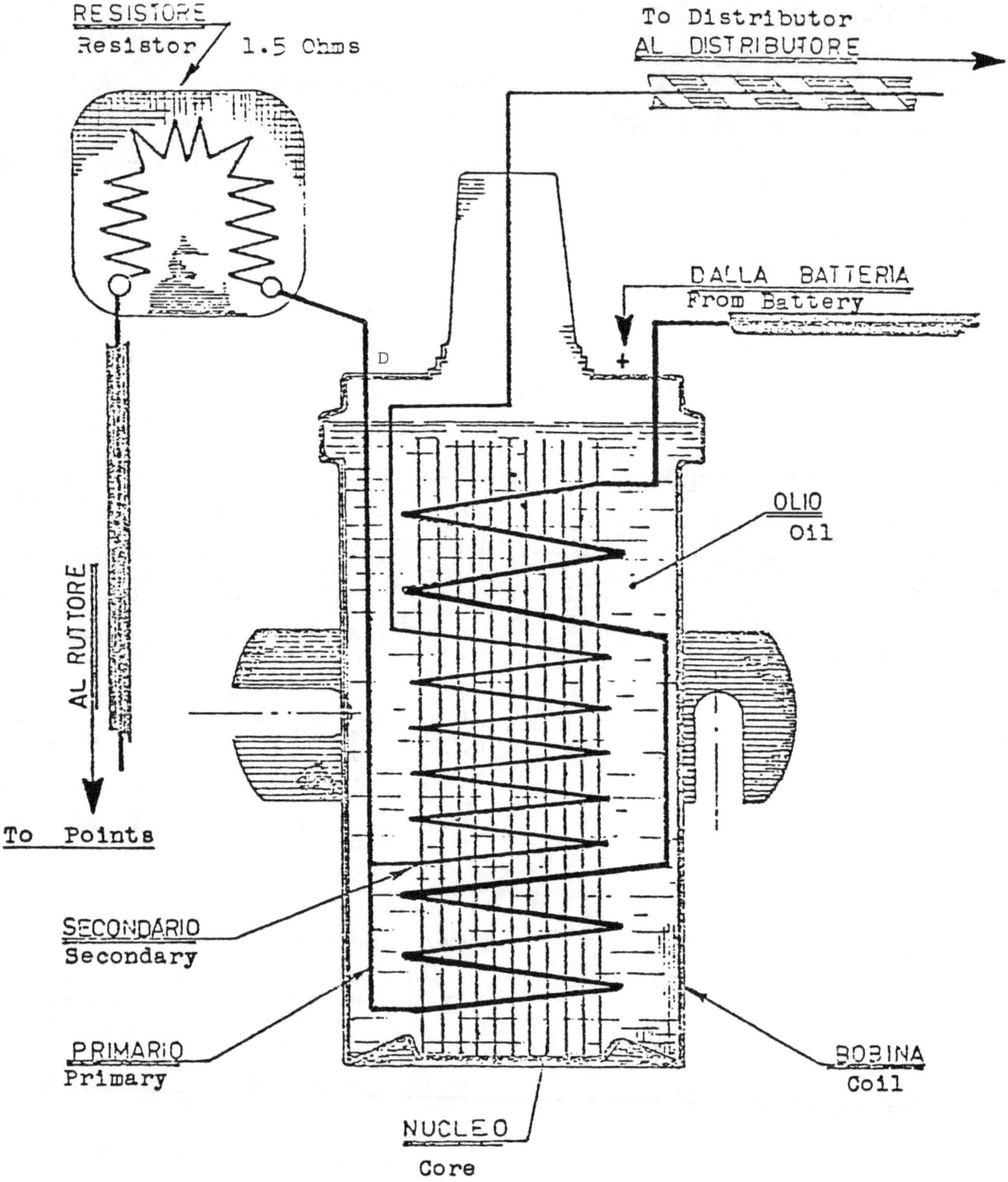

RESISTORE
Resistor  1.5 Ohms

To Distributor
AL DISTRIBUTORE

DALLA BATTERIA
From Battery

AL RUTTORE
To Points

OLIO
Oil

SECONDARIO
Secondary

PRIMARIO
Primary

NUCLEO
Core

BOBINA
Coil

Coil Primary Current, 3A at 13V.  High Voltage, 21KV Open Circuit.

## IDENTIFICATION OF FUSE BOX CIRCUITS

| LEFT FUSE BOX MARKINGS | FUSE SIZE | POSITION | CIRCUIT |
|---|---|---|---|
| Spineterogeni | 16A | 1 | Ignition Coils |
| Avviamento | 8A | 2 | Starter Solenoid |
| Avvis. - Accendisigari, Servizi | 16A | 3 | Horn, Cigar Lighter, Acces. |
| Spiadinamo - Ind. Livello, Overdrive - Ventilator | 8A | 4 | Gen. Lamp, Fuel Level, Overdrive, Fan, Instruments |
| Tergicristallo - Electroflux, Illuminazone Quadro - Servizi | 16A | 5 | Windshield Wipers, Electric Fuel Pump, Dash Lights, Acces. |
| Indicatori Direz - Stop Condizionatore | 16A | 6 | Stop and Turn Signals, Air Conditioner |

| RIGHT FUSE BOX MARKINGS | FUSE SIZE | POSITION | CIRCUIT |
|---|---|---|---|
| Abbagliante - Dex. | 8A | 1 | Right High Beam Lamp |
| Abbagliante - Sin. | 8A | 2 | Left High Beam Lamp |
| Antiabbagliante - Dex. | 8A | 3 | Right Low Beam Lamp |
| Antiabbagliante - Sin. | 8A | 4 | Left Low Beam Lamp |
| Posizione - Targa, Cofano - Retromarcia | 8A | 5 | Side Lamps, License Plate Lamp, Engine Comp. Lamp, Trunk & Back-up Lamps |
| Antinebbia, Plafoniere | 8A | 6 | Fog Lamps, Interior Lamps. |

## 250 GT SCHEMATIC WIRING DIAGRAM

1. Head lamps
2. Turn signals - front
3. Driving or fog lamps
4. Generator
5. Starter motor and solenoid
6. 12 Volt battery
7. Horn
8. Ignition coils
9. Distributor
10. Spark plugs
11. Under hood lamps
12. Fuses
13. Horn relay
14. Voltage regulator
15. Electrical panel
16. Junction blocks
17. Windshield wiper motor
18. Instrument lamps
19. Instrument regulator
20. Fuel level gauge
21. Turn signal relay
22. Stop light switch
23. Junction blocks
24. Head lamp relay
25. Horn button
26. Turn signal switch
27. Interior light switch
28. Electric fuel pump switch
29. High beam headlight switch
30. Interior light door switch
31. Turn signal indicating lamp
32. Generator warning lamp
33. Interior lamp
34. Instrument lamp switch
35. Windshield wiper motor
36. Fuel pump indicator lamp
37. Headlamp indicator lamp
38. Defroster indicator lamp
39. Blower switch
40. Blower
41. Headlight switch
42. Ignition switch
43. Electric fuel pump
44. Fuel gauge Float
45. Turn signals - rear
46. Stop and tail lamps

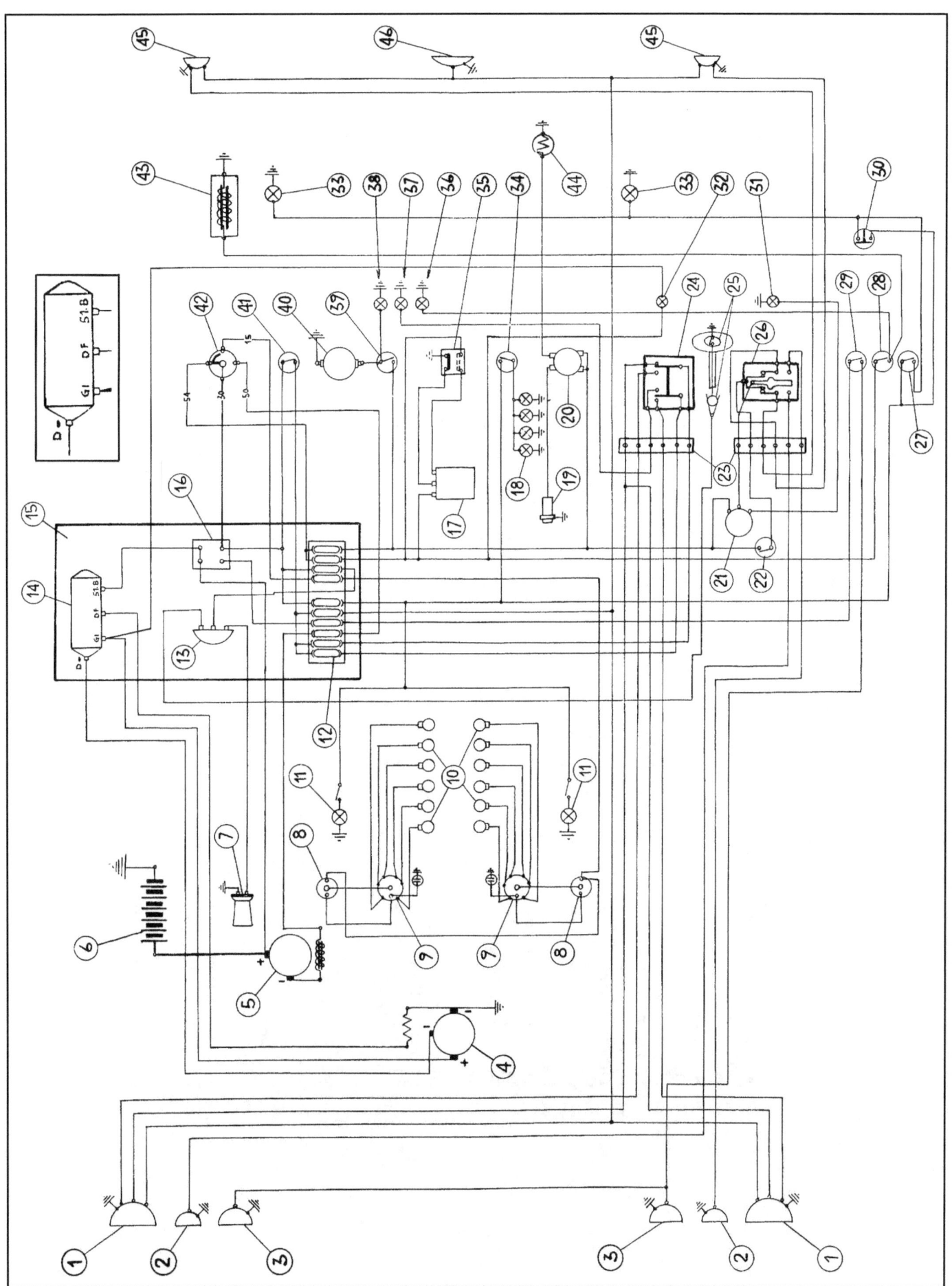

## 250 GT COMPONENT WIRING DIAGRAM

1) Headlamps and dip
2) Front lights and indicators
3) Fog lights
4) Dynamo
5) Starter
6) Battery
7) Horn
8) Horn relays
9) Ignition coils
10) Distributor
11) Spark plugs
12) Bonnet light
13) Panel
14) Installation fuses
15) Dynamo regulator
16) Terminal board
17) Windscreen wiper motor (2 speeds)
18) Switch for a/m. motor
19) Panel lights
20) Rheostat for panel
21) Thermocontact driving the fan
22) Fan for radiator
23) Petrol gauge
24) Tank level
25) Direction indicator relays
26) Commutator with direction indicator
27) Electric pump
28) Electric pump switch
29) Fog light switch
30) Hydraulic switch for stop lights
31) Outside light commutator
32) Horn push button
33) Inside light switch
34) Automatic switch for inside lights
35) Direction indicator lights
36) Dynamo charge lights
37) Inside lights lamp
38) Indicating lamp for electric pump
39) Heater fan indicating lamp
40) Indicating lamp for headlights
41) Conditioner switches
42) Conditioner electric fans
43) Deviolux (Front light commutator relay)
44) Ignition switch
45) Rear lights
46) Number plate light
47) Reverse lamp
48) Reverse light switch (on the gearbox)
49) Solenoid Overdrive
50) Switch on gearbox driving the Overdrive
51) Switch below the steering for Overdrive
52) Relay driving the Overdrive
53) Boot light
54) Water and oil thermometers
55) Side direction indicators (arrows)
56) Lighter
57) Signaling light for fuel level
58) Side door lamps
59) Thermocontacts for water and oil thermometers

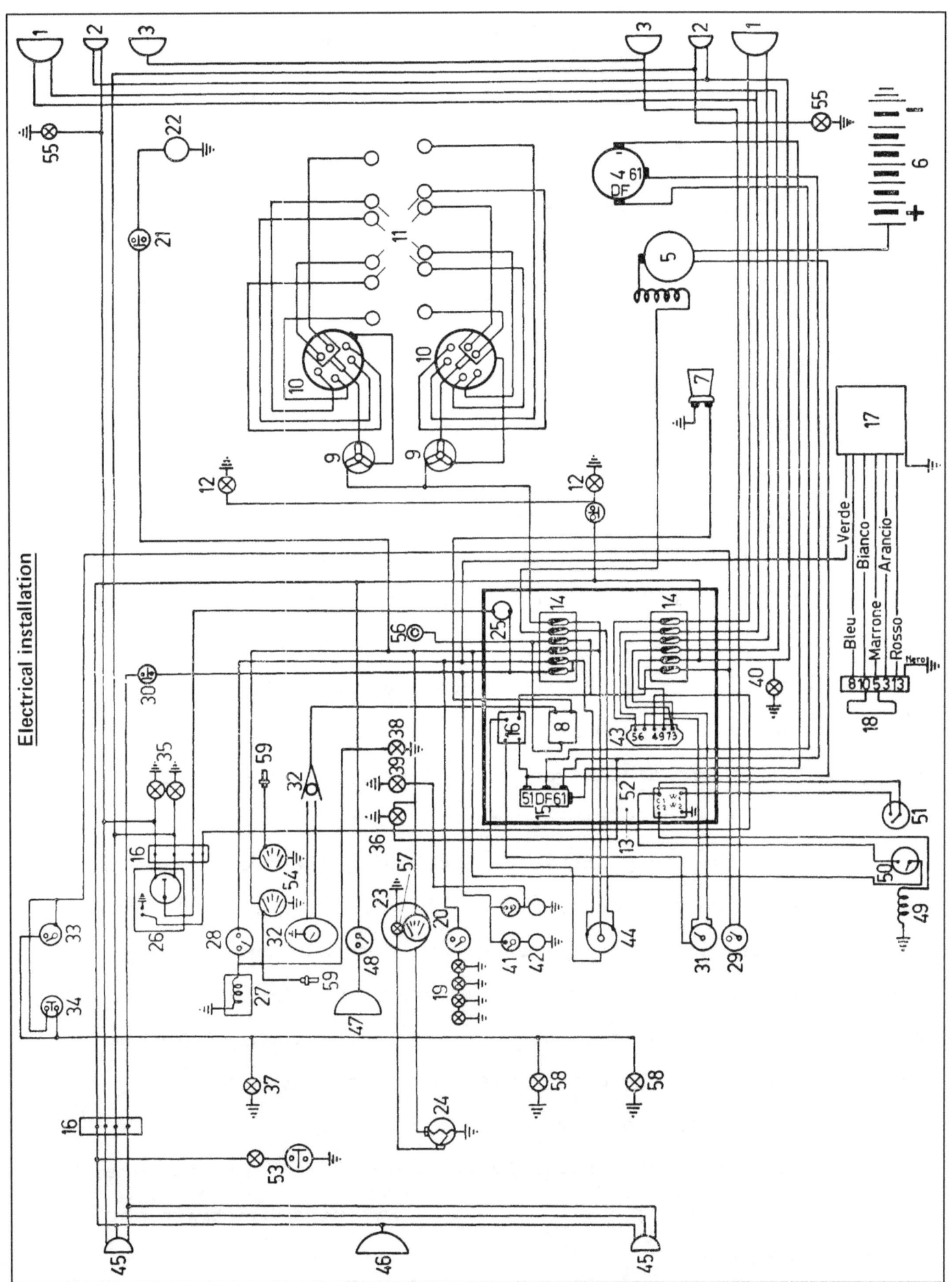

# FERRARI DISTRIBUTOR TIMING

To completely time a late model Ferrari, a timing light (electronic) and a feeler guage are required.

First install all new point sets, or file (with an ignition point file) the old sets so that a smooth parallel surface remains. Then adjust the gaps to the spacing recommended in the owners manual. (.014") Connect a good quality timing light to the #1 spark plug. Start the engine and let it idle below 1000 RPM. The 10AF mark, which is just before the PM1/6 on the flywheel, should be visible at engine idle as observed in the timing window. Now increase the RPM to 5000; 42AM should be visible on the flywheel. If the 10AF lines up with the pointer and the 42AM does not, advance the right distributor (by rotating the top) until the 42AM is at the pointer. It is generally preferred that the engine be in time at high speed more so than at idle. Next move the timing light to sparkplug #6 on the same bank. Start the engine and increase the RPM to 5000. The same 42AM (now 360° later) should be seen at the pointer. If it is not visible, stop the engine and adjust the position of the second point set by loosening the two screws and sliding the set forward or backward, thus the timing of the even number cylinder can be accomplished. To determine which points control the even and odd number cylinder, crank the engine over until the rotor points to #1 on the cap. The set of points that are just opening are the even numbers set (the flywheel mark should be just coming up to the 10AF mark also). This "odd set" is not adjusted for position, only the "even number set" is moved to adjust the #6 firing position. This completes the timing of the right hand bank.

# FERRARI DISTRIBUTOR TIMING

To time the left hand bank, connect the timing light to cylinder #7 (closest to the fire wall) and start the engine. At an idle (less than 1000 RPM), the 10AF mark should be visible at the pointer. This mark is located just ahead of the PM7/12 mark on the flywheel. Increase the engine speed to 5000 RPM and observe the timing mark; 42AM should be visible at the pointer. If this mark is not at the pointer, adjust the distributor (by rotating the top) so 42AM is at the pointer. Again, it is more important that the timing be correct at 42AM than at 10AF. Move the timing light to cylinder number 12. Start the engine and increase the RPM to 5000. The 42AM should again be visible at the pointer. If it is not, move the second or odd number points until the 42AM is obtained. To determine even #12 from the odd #7 point set, crank the engine until the rotor points to #7 on the distributor cap. The points should be just opening (10AM should be close to the pointer also); this set is the even number and should not be moved for cylinder number 12 adjustments.

It is always wise to repeat all timing checks again to double check their accuracy.

NOTES:

1. Be sure to retighten all of the 14mm nuts that hold the distributor top on to the base. These nuts should be just loose enough to move the distributor timing when adjusting #1 and #7 maximum advances (42AM).

FERRARI DISTRIBUTOR TIMING

NOTES (continued)

2. To avoid the sparking between the top and base of the distributor while timing, a short length of wire should be permanently connected to the top section. This cable should then be connected to the fire wall at some convenient point. Thus a good electrical connection between the distributor and ground is always maintained. A typical connection would be between the capacitor mounting screw and the ignition coils mounting bracket. A number 18 guage insulated wire is sufficient.

3. It is not always necessary to remove the metal cover plate on the bellhousing to expose the flywheel timing marks. A clear plastic (preferably Lexan) cover plate can be made by tracing the old piece and installing in its place. This way, the timing light can be shined through the "window" to check timing.

4. Lube the distributor cam lobes with a good quality grease such as Delco cam lube.

5. Definitions of the flywheel markings are:

   PM1/6  - Punto Morto; this is top dead center for #1 cylinder.
   PM7/12 - Punto Morto; this is top dead center for #7 cylinder.
   10AF   - 10° Anticipo Fillo; this is 10° fixed advanced.
   42AM   - 42° Anticipo Massimo; this is 42° maximum or full advance.

## Cam marks and pointer alighment

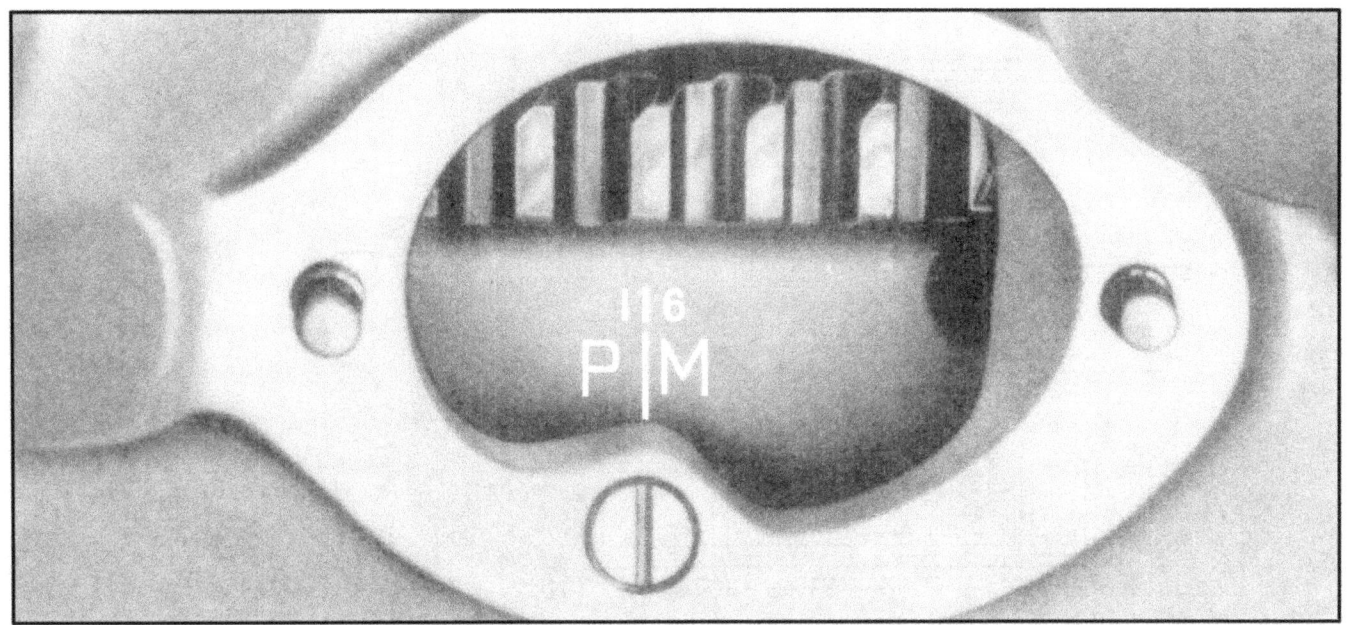

Fig. 1 — Segni di riferimento P.M. incisi sul volano. TDC for Cyl. #1

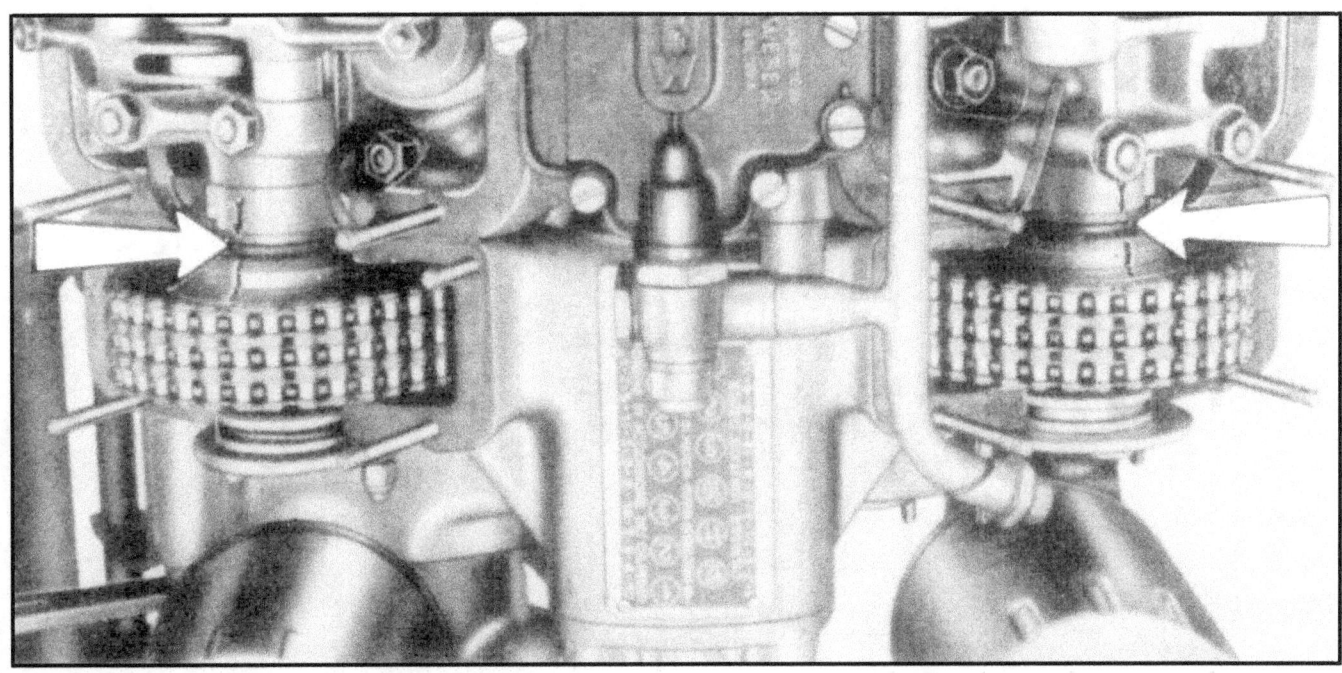

Fig. 2
Riferimenti incisi sugli alberi a camme e sui cappelli dei supporti bilancieri.

# IGNITION TIMING OF ENGINE

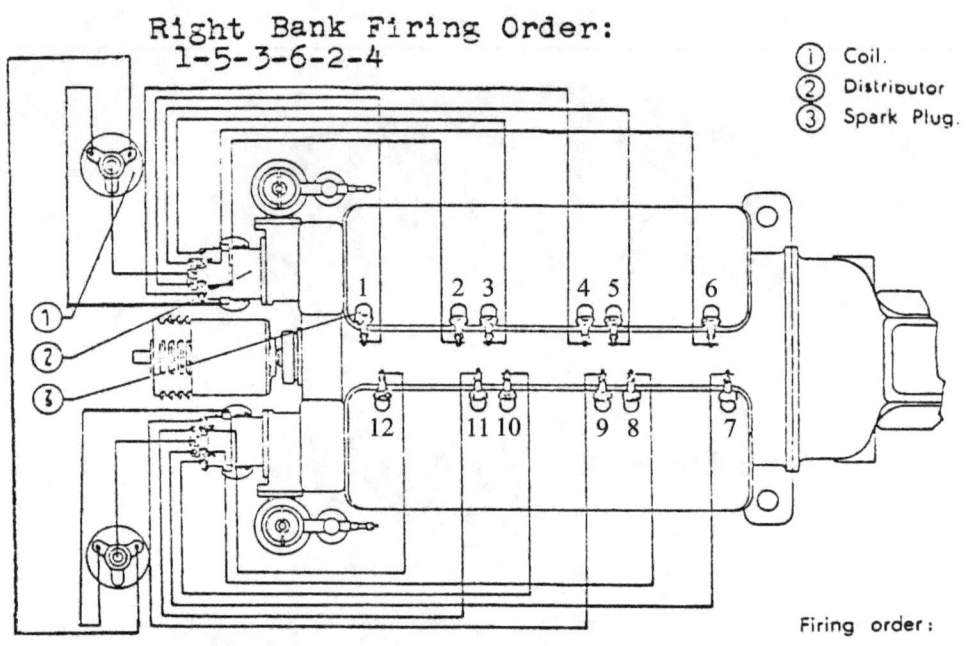

Right Bank Firing Order:
1-5-3-6-2-4

Left Bank Firing Order:
7-11-9-12-8-10

(1) Coil.
(2) Distributor
(3) Spark Plug.

Firing order:
1-7-5-11-3-9-6-12-2-8-4-10

Cylinder ignition order:
1-7-5-11-3-9-6-12-2-8-4-10

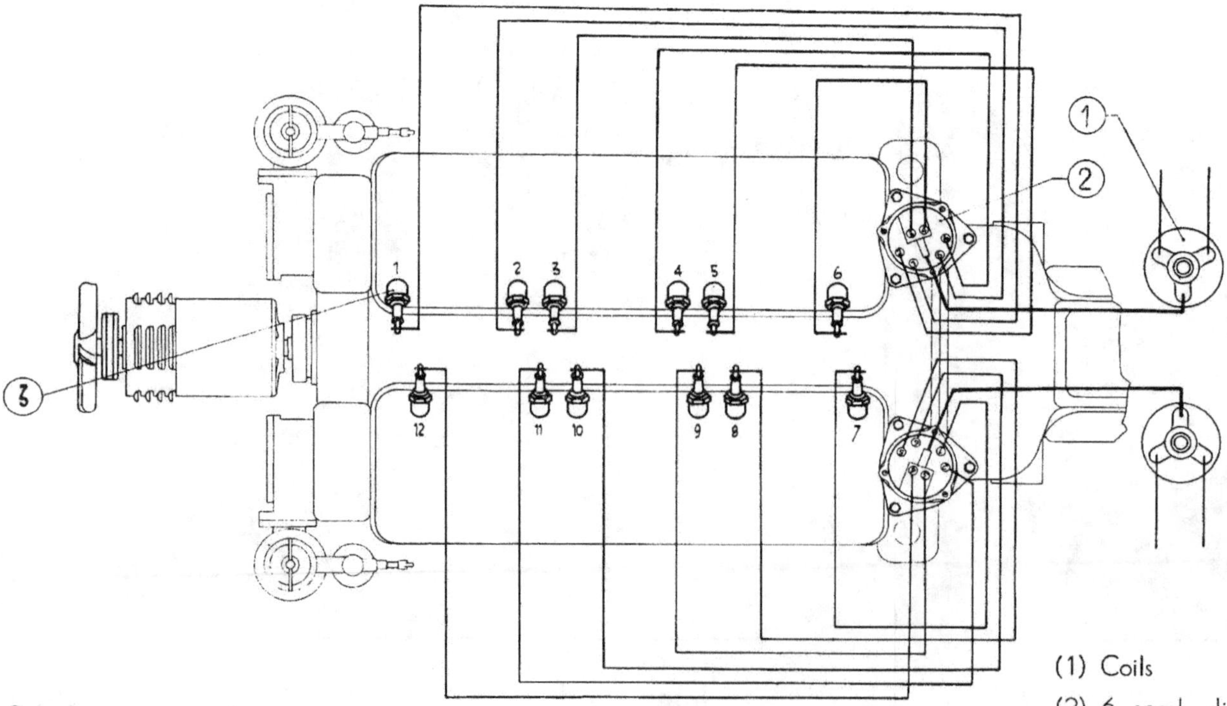

(1) Coils
(2) 6 spark distributors
(3) Spark plugs

# DISTRIBUTOR

Description: Magneti Marelli S85A Series.

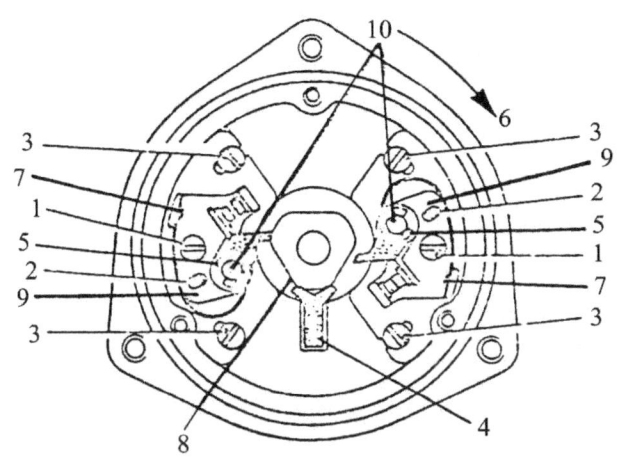
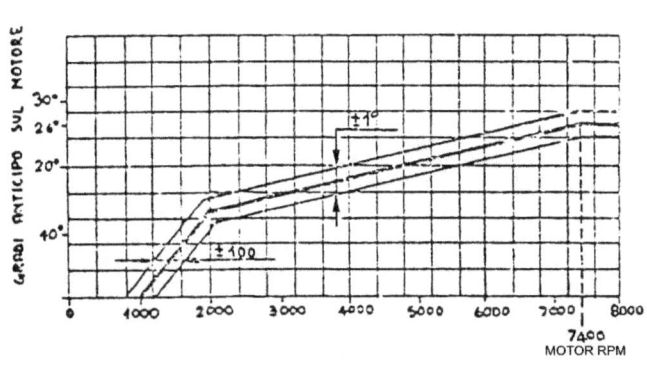

1. Gap Adjusting Screw
2. Pivot for Gap Adj.
3. Syncronizing Screw
4. Lube Wiper
5. Circ Ring for Points Removal
6. Direction of Rotation
7. Connection to Coil
8. Cam
9. Replaceable Point Assembly
10. Stationary Post
11. Rotor
12. Capacitor
13. Bearing Support
14. Base

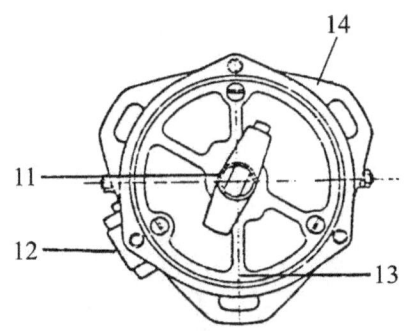
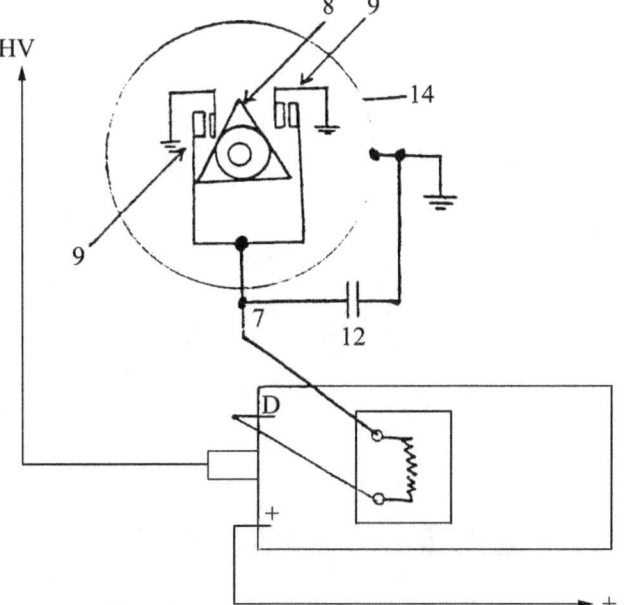

Point Gap      .014"

Capacitor      .18MFD

Points         4.71.0071.02

29

## DISTRIBUTOR

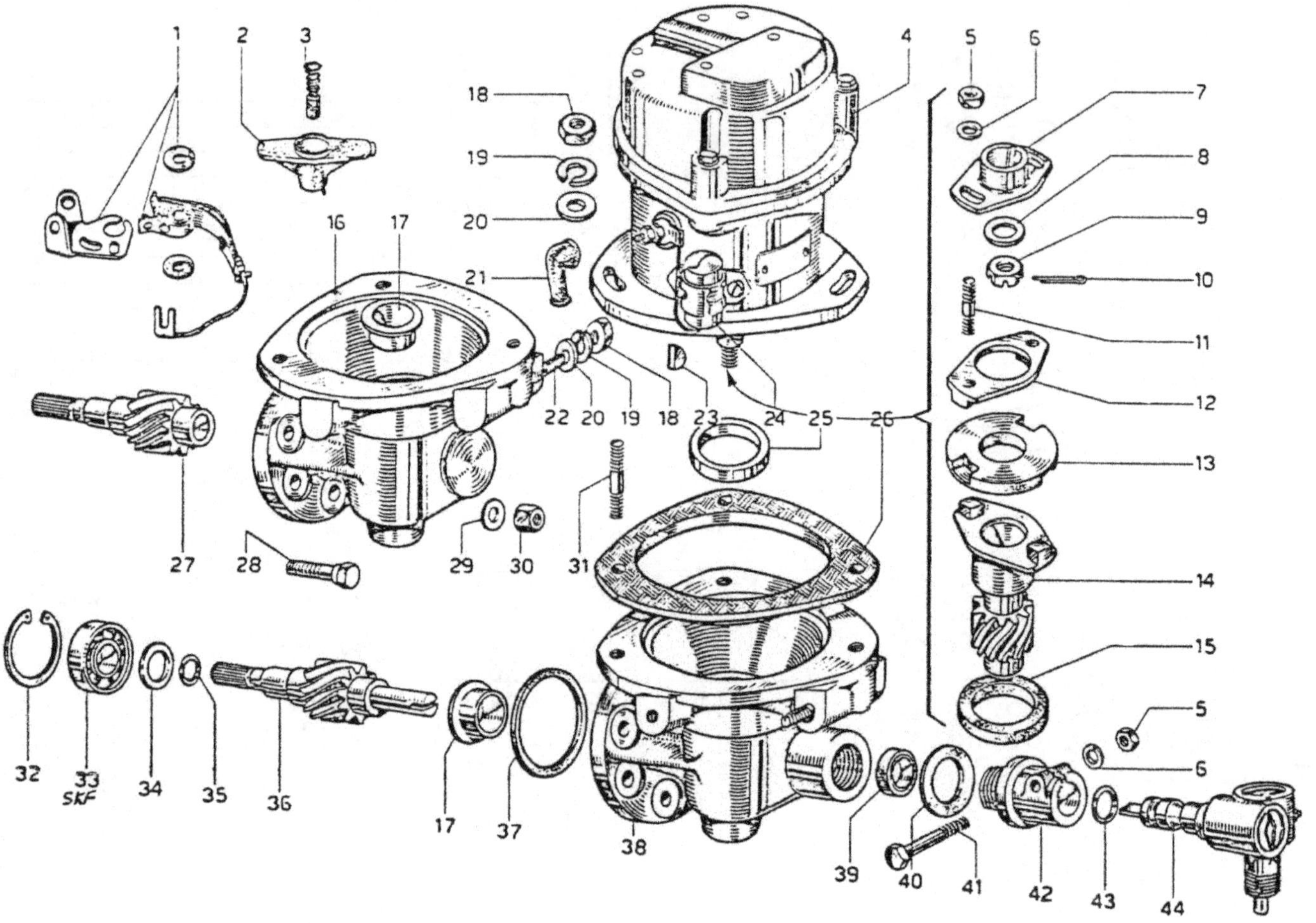

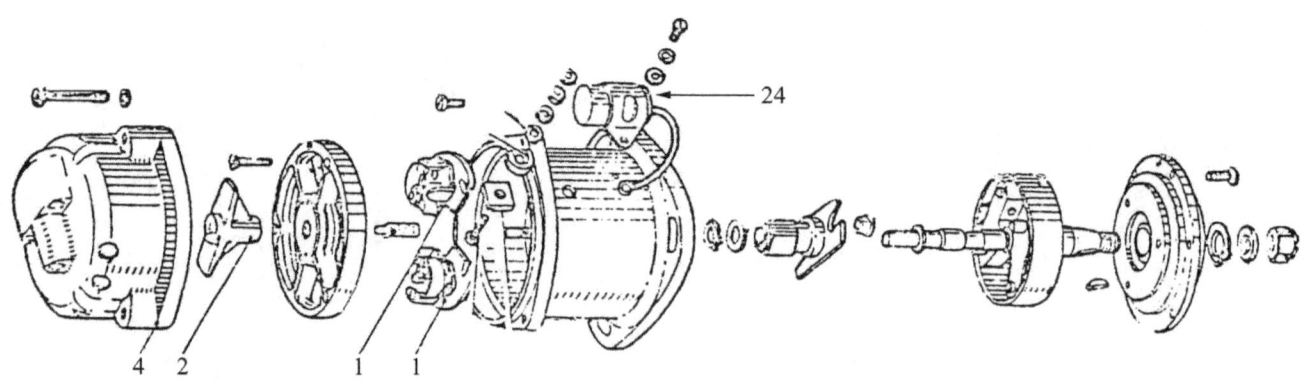

| NUMBER | DESCRIPTION | PART NUMBER | MANUFACTURER |
|--------|-------------|-------------|--------------|
| 1 | Point Assembly | H.71.0071.02 | Marelli |
| 2 | Rotor | 703.880.01 | Marelli |
| 3 | Wiper Brush | 91.006.01 | Marelli |
| 4 | Cap | 703.884.01 | Marelli |
| 24 | Capacitor | CE1E.18MFD | Marelli |
| 44 | Tach Drive | 50.0419.990.0 | Borletti |

## MARELLI DRAWING #5: ST 100 DTEM A,B,C,D,E   #6: ST 195 DTEM-KS

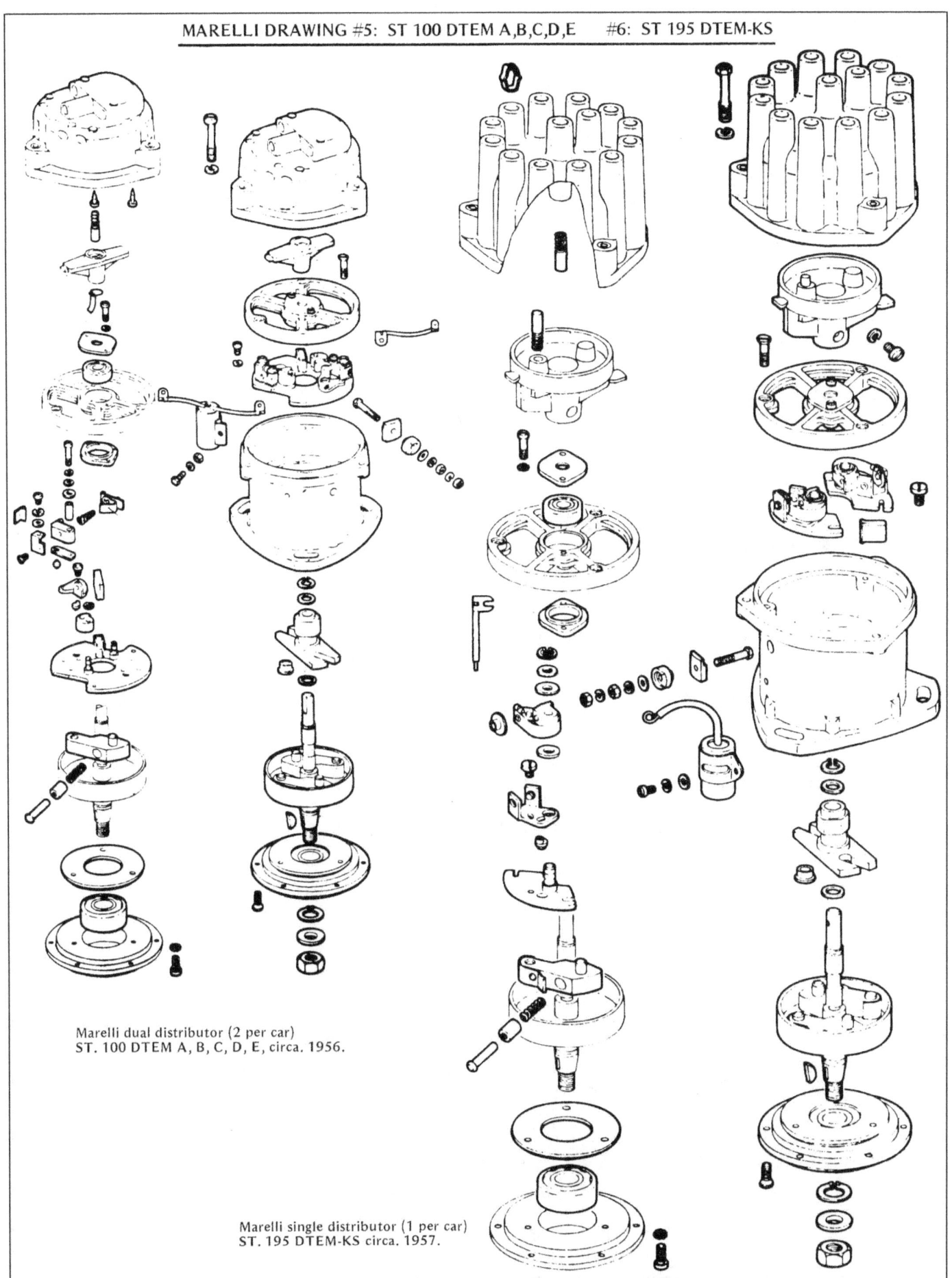

Marelli dual distributor (2 per car)
ST. 100 DTEM A, B, C, D, E, circa. 1956.

Marelli single distributor (1 per car)
ST. 195 DTEM-KS circa. 1957.

## IGNITION TIMING CHECK

To check the ignition timing on 250-275-330 engines, the following steps are required:

1. Remove the inspection plate on the top of the bellhousing.

2. Clean all markings on the flywheel with solvent.

3. Connect an electronic timing light (strobe) to cylinder #1 on the right bank.

4. Start the engine and allow it to idle below 1000 RPM.

5. Aim the timing lamp into the inspection port above the flywheel; 10AF should be near the fixed pointer.

6. Increase the RPM to 4500 RPM, 42AM should be indicated at the pointer.

7. Stop the engine and connect the timing light to cylinder #7 on the left bank.

8. Start the engine and let idle below 1000 RPM.

9. Aim the timing light into the inspection port above the flywheel; 10AF should be near the fixed pointer.

10. Increase the RPM to 4500 RPM: 42AM should be indicated at the pointer.

11. If any adjustment is required, loosen the three bolts on the distributor and rotate top until timing is correct.

12. Tighten all distributor nuts, and replace the inspection port cover.

NOTE: Ignition points should be clean and adjusted to .014" gap.

## MARELLI DRAWING #1: DN 22 A,B,C   #2: DN 25 A   #3: DN 2 G

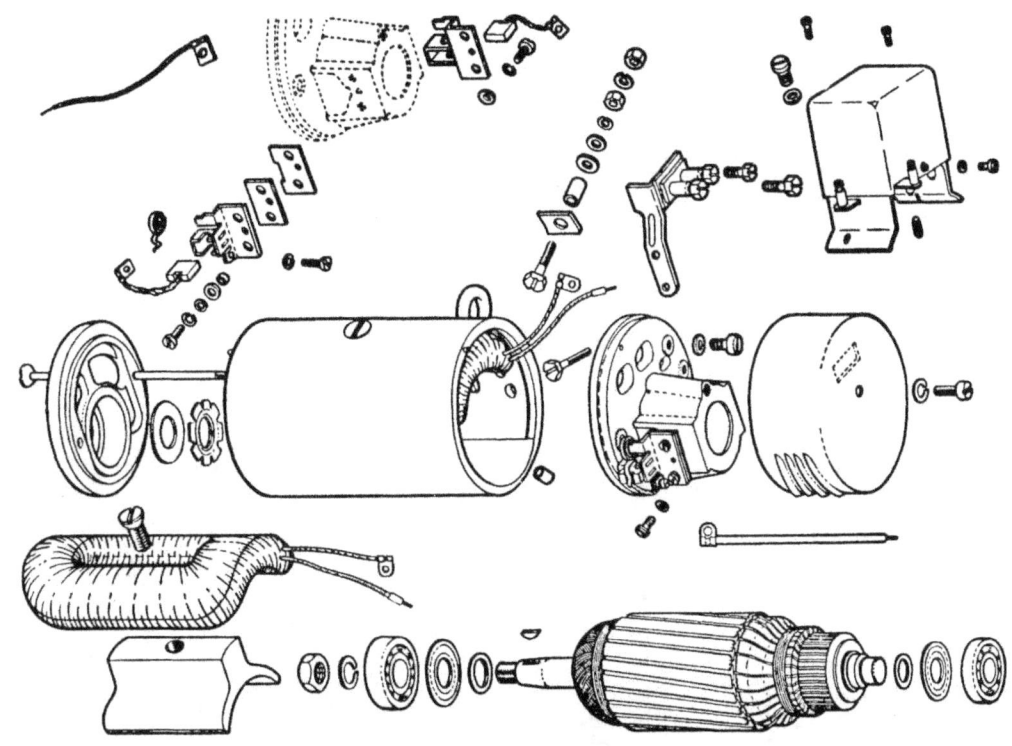

Marelli DN 22 A, B, C, DN 25 A Generator with regulator at top right.

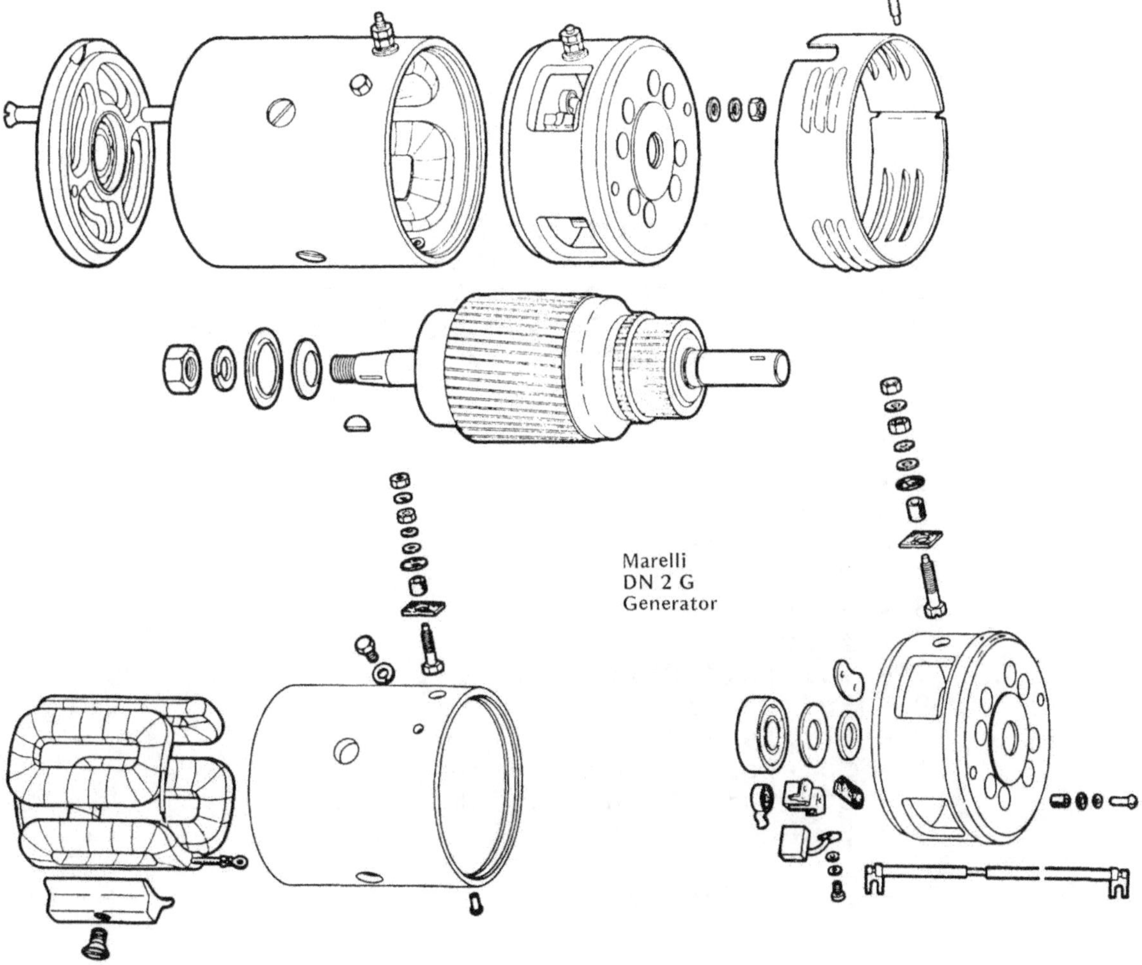

Marelli DN 2 G Generator

33

# STARTER MOTOR

DESCRIPTION: Magneti Marelli 12 Volt Cranking Motor MT Series.

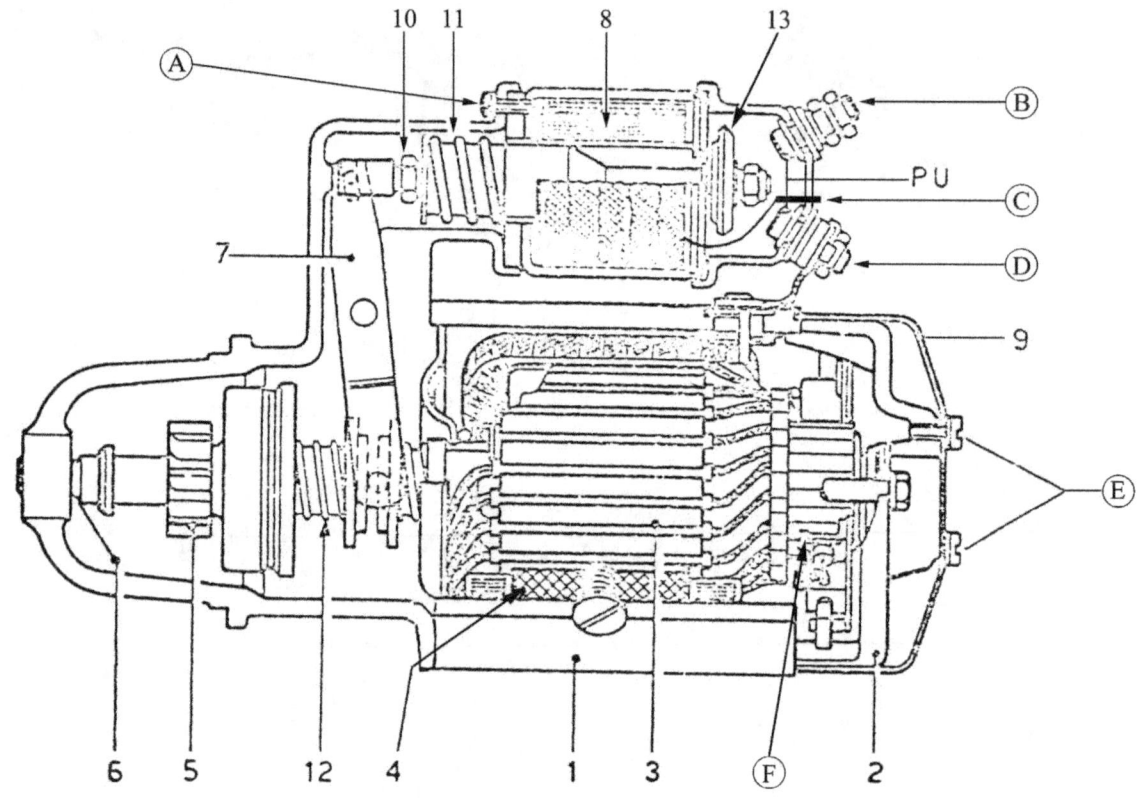

```
A - Solenoid Mounting through Bolts
B - Battery Positive Cable
C - Ignition Switch Start Terminal
D - Solenoid to Starter Motor Terminal
E - End Cover Securing Screws
F - Communtator

1 - Case or Frame
2 - Rear Bearing Support
3 - Armature
4 - Field Winding
5 - Drive Gear
6 - Front Bearing
7 - Actuating Link
8 - Solenoid Coil
9 - End Cover
10 - Stroke Adjustment
11 - Solenoid Disengagement Spring
12 - Starter Disengagement Spring
13 - Solenoid Contactor

PU - Fixed Contacts
```

# MARELLI DRAWING #4: MT 21 A,B,C,D,E,F,G,H,L,M,N,P,Q,R,S

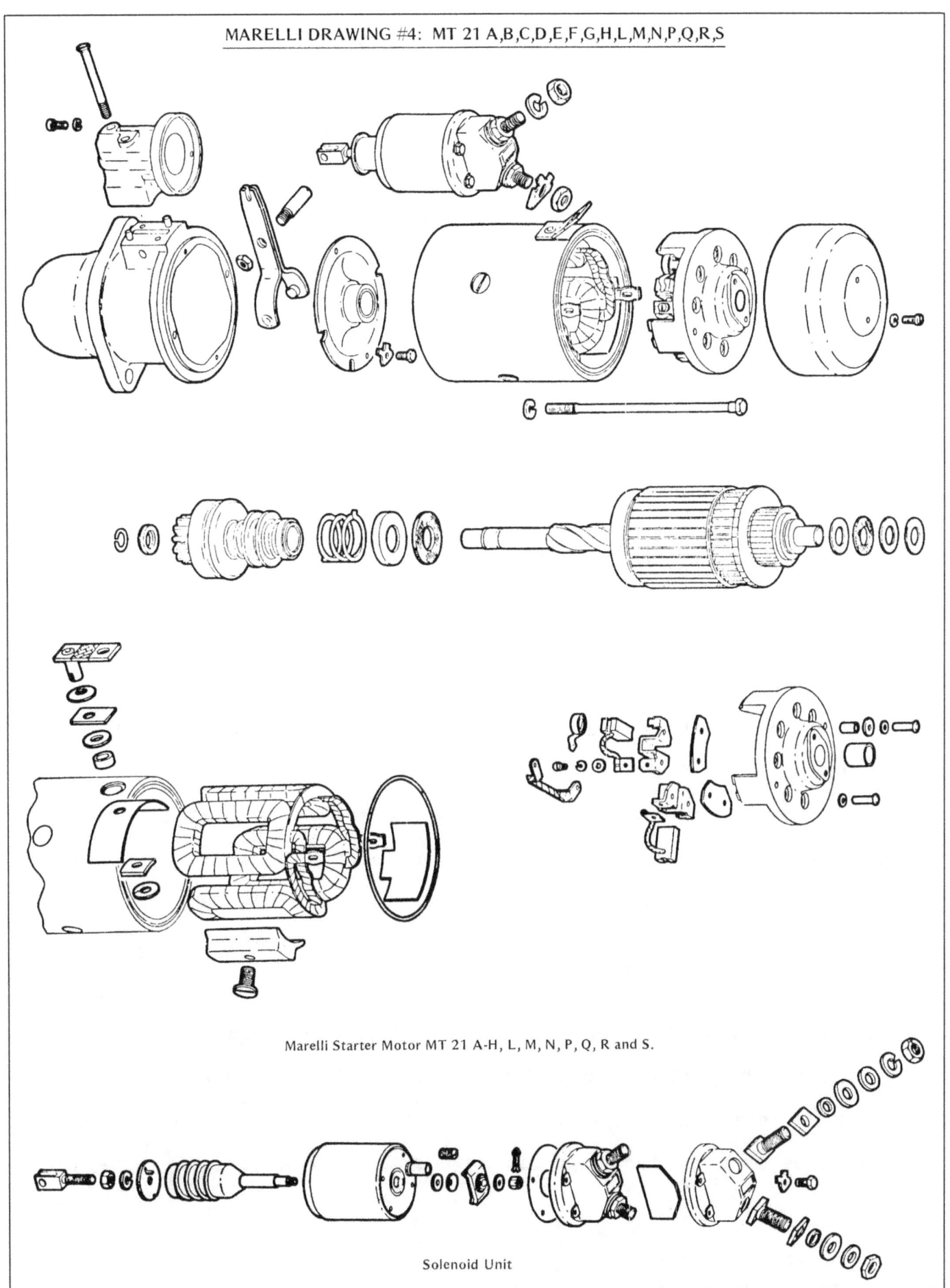

Marelli Starter Motor MT 21 A-H, L, M, N, P, Q, R and S.

Solenoid Unit

# STARTER MOTOR

To service a defective cranking motor, the following steps are suggested:

## Removal

1. Remove the battery cables from the battery.
2. Remove the battery if it is on the starter motor side.
3. Remove all bolts at muffler-header connection.
4. Remove all header nuts and lock washers.
5. Carefully manipulate the exhaust header assembly out of the engine compartment.
6. Pull the clad gaskets off of the head.
7. Remove all cables from the starter motor.
8. Remove the two starter motor mounting nuts and lockwashers. (If the crosmember interferes with the removal of the mounting nuts, loosen the engine mounts and jack the engine up slightly.)
9. Carefully slide the startermotor out of the bellhousing.

## Testing

Holding the starter motor securely on a bench, connect the negative cable of a fully charged 12 Volt battery to the case (1) on the motor and the positive lead to the solenoid terminal (B) where the heavy (vehicle) cable was connected. By temporarily connecting a test wire between the small (C) terminal of the solenoid and the positive (B) lead to the battery, the solenoid should engage and the motor begin to turn. If the solenoid does not produce a definite clunk, indicating engagement, it may be defective. Separate the solenoid from the starter by removing the two screws (A) securing it to the frame, and disconnecting the wire (D) to the motor. The coil can be removed from the solenoid and the contacts inspected.

## STARTER MOTOR (continued)

If they are burned or dirty, file them flat and smooth. For severely damaged contacts, machine flat or replace. The contactor, 13, should be cleaned smooth. Inspect the coil for burnt wire or water damage. Re-assemble the solenoid into the housing, and tighten the through bolts (A) securely. The coil should energize when battery voltage is applied to the small terminal (C). A definite klunk should be heard and the engagement gear, 5, should move forward. Connect the lead from the starter motor to terminal (D), and repeat the test. With battery voltage applied to terminal (B) and a jumper wire connected from (B) to (C) the motor should spin rapidly, about 200-400 RPM. If the motor appears sluggish in performance, then an overhaul is indicated. Remove the two end cap screws (E), and pull the end cover, 9, off. Inspect the brushes and commutator for damage or wear. Replace worn brushes with new ones obtained from a dealer. Smooth the commutator (copper colored area) (F), with #400 sandpaper. Clean the slots between each segment of the commutator with a thin blade. Clean and lube the bearings on each end. Springs 11, and 12 should be in good condition, replace any broken or weak disengagement springs with new parts. Never use solvents or thinners to clean the inside of the starter, improper chemicals may soften the insulation on the windings. If the starter draws heavy current and still does not spin fast, the windings may be shorted. Remove the armature (3) and have it tested on a "growler" to find any damage. Replacement windings can be ordered through authorized dealers.

NOTES

# 250GT ~ CARBURETTORS

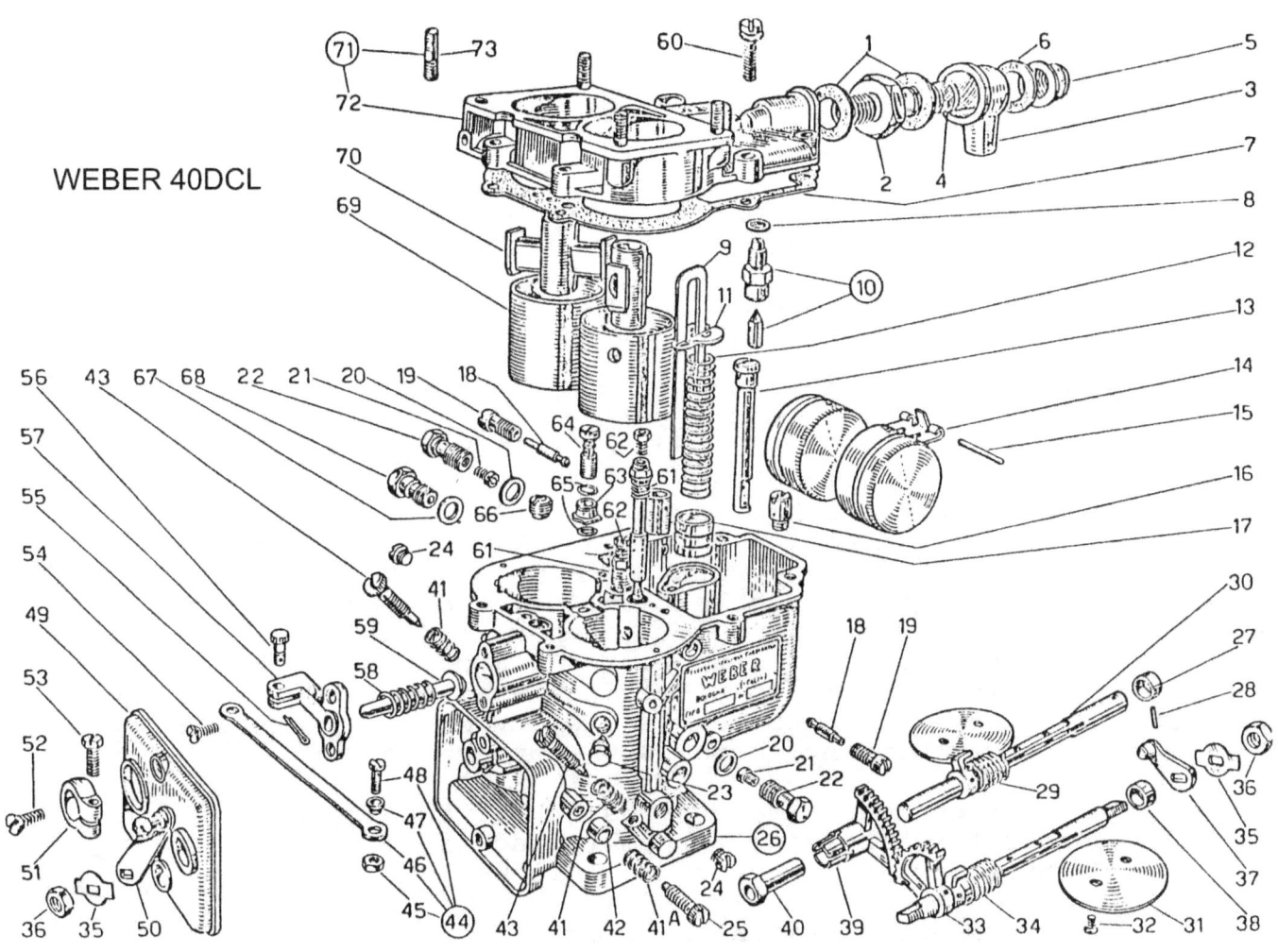

WEBER 40DCL

| | |
|---|---|
| ADJUSTMENT | 40 - 42 |
| SETTINGS | 43 |
| WEBER 40 DCL (Ferrari Factory Documentation) | 44 - 46 |
| WEBER 36-40-42 DCF/DCL/DCZ (Weber Factory Documentation) | 47 - 48 |
| WEBER 36 DCS (Ferrari Factory Documentation) | 49 - 50 |
| WEBER 36 DCS (Weber Factory Documentation) | 51 - 54 |

## CARBURETOR ADJUSTMENT

The carburetor tuning on most late model Ferraris can be accomplished without a great deal of effort.

PREREQUISITES:

Ignition timing must be correct. Spark plugs should be in good condition or new. Engine should be warm or hot - 140°F oil temp. min. Air cleaners should be clean or new. Weather should be good - no rain or snow. Engine should be run hard prior to tuning - to blow out carbon, etc. Valve clearances should be set and correct. Exhaust system free from holes or leaks. Fuel filters clean. Float levels set to specification.

EQUIPMENT REQUIRED:

10mm wrench.
Screwdriver with 1/4" blade.
Electronic engine tachometer
Uni-syn or equivalent.

If all of the prerequisites have been followed, the next step is to adjust the idle mixtures and throttle opening settings. As previously stated, this procedure is for obtaining a smooth idle only. High speed mixture problems or float settings are not dealt with. If your Ferrari is operating to your satisfaction don't disturb it. If it does need adjustment, here is the procedure:

1. Close the choke (keeps parts out of the engine).
2. Remove the aircleaner assembly and filters.
3. Carburetors are numbered as follows:

    #1 - closest to radiator (front)
    #2 - center
    #3 - closest to fire wall (rear)

4. Remove the linkage clips from carburetors #2 and #3, and lift the linkage rods away from the carburetors.
5. Open the chokes and start the engine (still at warm operating temperature) and leave it idle.
6. Increase the engine RPM to approximately 1000 RPM with carburetor #1's throttle adjustment screw.
7. Completely close (CCW) the throttle adjustment screws on carburetors #2 and #3.

8.  The engine's rpm should be adjusted with carburetor #1 throttle screw so an idle of about 800 RPM can be obtained.
9.  Connect an electronic tachometer to either distributor.

    NOTE: The RPM readings may not be correct, but we are only interested in meter movement, not readings.

10. Adjust both of carburetor #1's idle mixture screws until a maximum RPM indication is obtained on the electronic tachometer. (Engine is running rough because only one carburetor is in operation)
11. Increase the engine RPMs (3000-4000) by blipping (pressing) the linkage of carburetor #1 with your hand. There should be no spitting (backfireing) through the carburetor or crackling at the exhaust when released. If so, repeat step ten with greater accuracy.
12. Increase the RPM of the engine via the throttle adjustment screw on carburetor #2, 1000 RPM is o.k.
13. Close the throttle adjustment screw on carburetor #1. #1 and #3 are now closed.
14. Adjust the throttle adjustment screw on carburetor #2 so that an idle of about 800 RPM is obtained.
15. Adjust both of carburetor #2's idle mixture screws until the maximum RPM indication is obtained on the electronic tachometer. Engine will continue to run rough because only one carburetor is operating the engine.
16. Increase the engine's RPMs (3000-4000) by blipping (pressing) the throttle linkage of carburetor #2 with your hand. There should be no spitting (backfireing through the carburetor or crackling at the exhaust when the linkage is released. If so, repeat step fifteen with greater accuracy.
17. Increase the RPM of the engine via the throttle adjustment screw on carburetor #3, 1000 RPM is o.k.
18. Close the throttle adjustment screw on carburetor #2; #1 and #2 are now closed.
19. Adjust the throttle adjustment screw on carburetor #3 so that an idle of about 800 RPM is obtained.
20. Adjust both of carburetor #3's idle mixture screws unitl the maximum RPM is obtained on the electronic tachometer. The engine will continue to run rough, because only one carburetor is operating the engine.
21. Increase the engine RPMs (3000-4000) by blipping (pressing) the throttle linkage of carburetor #3 with your hand. There should be no spitting (backfireing through the carburetor) or crackling at the exhaust when the linkage is released. If so repeat step twenty — with greater accuracy.

    NOTE: Exhaust leaks will cause crackling at the exhaust as though the mixture were too lean. All holes should be plugged as tight as possible; try muffler cement as a quick fix.

22. Balancing of the throttle openings. With the engine still idling, turn each of the throttle adjustment screws in (CW) until each carburetor contributes to an increase in engine RPM. (NOTE: Linkages are still disconnected, and the engine warm).
23. Place the UNI-SYN instrument on top of carburetor number three. Adjust the UNI-SYN until the cork is at a readable line. (NOTE: On carburetors that have chokes that prevent the placement of the UNI-SYN atop of the venturi, a small plastic or sheetmetal adaptor will need to be fabricated to clear the choke butterflies.
24. Place the UNI-SYN on carburetor number one and adjust the throttle adjustment screw until the same reading (cork level) is indicated.
25. Place the UNI-SYN on carburetor #2 and adjust the throttle adjustment screw until the same reading (cork level) is indicated.
26. The engine should idle smooth at about 800 RPM. To properly adjust for this RPM, adjust the throttle adjustment screws on all carburetors until this RPM is indicated and the cork levels in the UNI-SYN are the same on all carburetor venturi.
27. Stop the engine. Loosen all of the adjustments of carburetor #2 and #3's throttle linkage rods so that they are free to be lengthened or shortened. Loosen the actuating lever arms that operate carburetors #2 and #3 so that they are free to move. (These are located on the main linkage bar).
28. Place the lever arms (that actuate the linkage rods) so that they are in line with the lever arm of carburetor #1's linkage arm. Tighten the bolts that secure these arms.
29. Adjust the lengths of the carburetor linkage rods until they slip loose over the ball on the lever arm. Secure the jam nuts, and replace the spring clips.
30. Start the engine, recheck the idle; it should steady at about 800 RPM on the engine's tachometer. If it does not, repeat steps 22 to 29.
31. Drive the vehicle in second gear, at about 3000 RPM, release the accelerator pedal; there should be no crackling heard at the exhaust as the engine slows down (in gear) to 1000 RPM. If crackling is heard, repeat steps 10, 15, 20 and 22 to 29.
32. Close the chokes, replace the aircleaner, filter elements and top plate. Open the choke (this is done to prevent any loose hardware from falling into the carburetor).

Some carburetors and adjustment details may differ greatly from the systems described (250GT and 330GT engines),

## CARBURETTOR SETTINGS

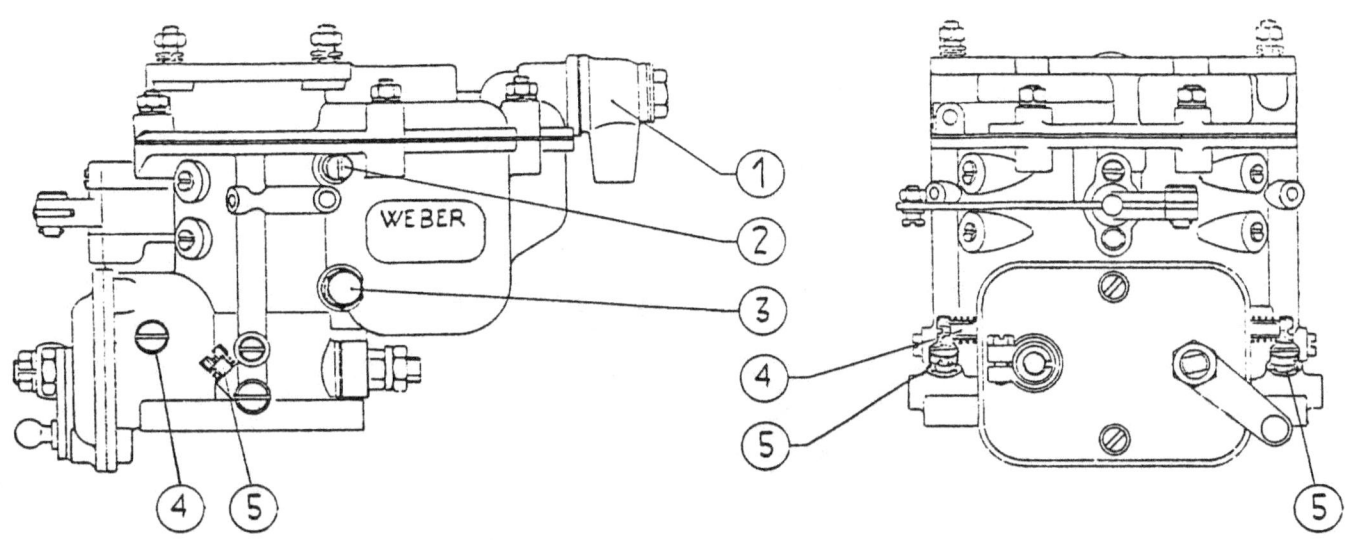

(1) Fuel filter
(2) Idling jet screw
(3) Main jet screw
(4) Throttle opening screw
(5) Idling mixture screw

| SETTING DETAILS | 330GT ADJUSTMENT | 275GT ADJUSTMENT | 250GT ADJUSTMENT |
|---|---|---|---|
| Choke | 27mm | 3.5mm | 27mm |
| Main jet | 1.30mm | 1.35mm | 1.50mm |
| Air Corrector jet | 1.80mm | 2.00mm | 1.80mm |
| Idling jet | 0.60mm | .60mm | .60mm |
| Idling air | 1.20mm | 1.25mm | 1.20mm |
| Idling jet holder | 3.50mm | - | 2.50mm |
| Pump jet | 0.60mm | .60mm | .60mm |
| Pump stroke | 4-4.50mm | 4.5mm | 3.00mm |
| Needle seat | 2.25mm | 1.75mm | 1.75mm |
| Float level | 3-3.25mm | 3-3.5mm | 3-3.25mm |

# CARBURATORI WEBER
## Tipo 40 DCL

**Applicazione**
FERRARI 250 GT

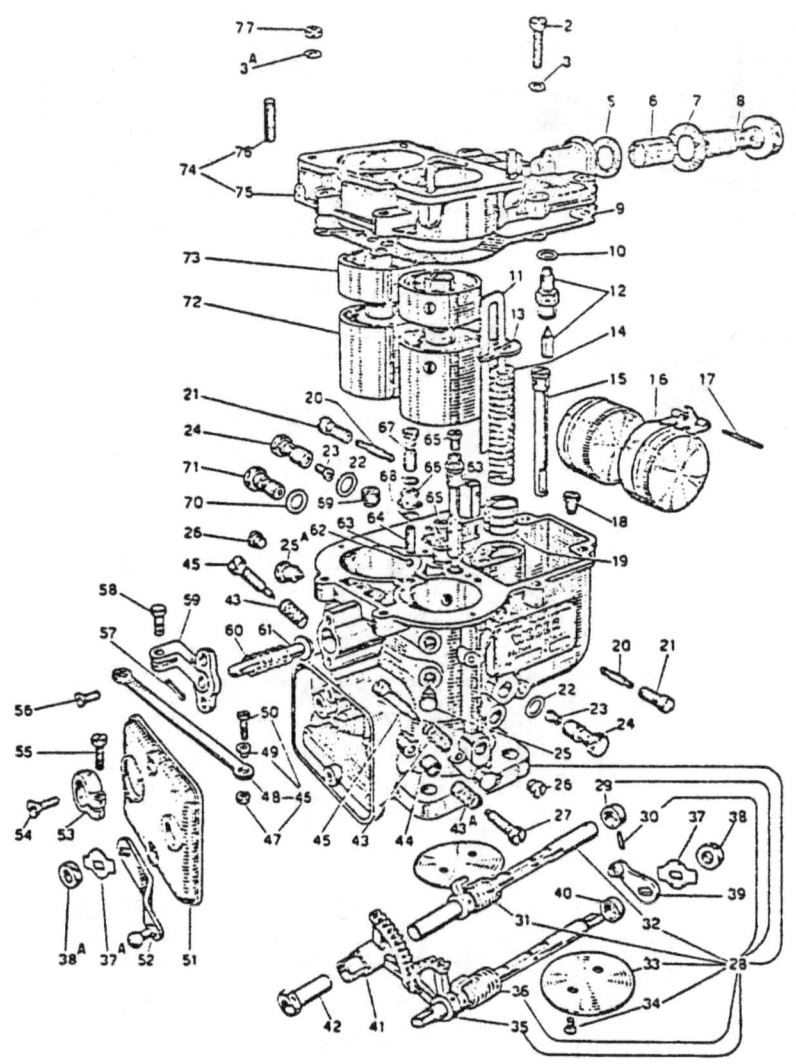

Nelle ordinazioni indicare: numero di matricola del particolare richiesto, la eventuale taratura e il numero e tipo del carburatore.

## (\*) NORME PER LA LIVELLATURA DEL GALLEGGIANTE

Per effettuare la livellatura del galleggiante è necessario attenersi alle seguenti norme di carattere generale:

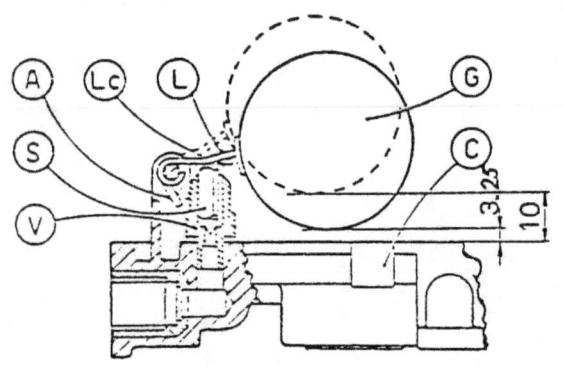

1) - accertarsi che la valvola a spillo (V) sia bene avvitata nel suo alloggiamento;

2) - con coperchio carburatore (C) capovolto, i due semigalleggianti devono distare dal piano del coperchio stesso mm. 3.25

3) - a livellatura effettuata, controllare che la corsa del galleggiante (G) sia di mm. 7 modificando eventualmente la posizione dell'appendice (A).

4) - Qualora il galleggiante (G) non fosse giustamente impostato, modificare la posizione delle linguette (L) del galleggiante stesso fino a raggiungere la quota richiesta, avendo cura che la linguetta (Lc) sia perpendicolare all'asse dello spillo (S) e che non presenti, sul piano di contatto, incaccature che possano influire sul libero scorrimento dello spillo stesso.

# CARBURATORI WEBER
## Tipo 40 DCL

**Applicazione** FERRARI 250 GT

## REGOLAZIONE

| Figura | Quantità | Numero di ordinazione WEBER | | DENOMINAZIONE | Taratura in mm |
|---|---|---|---|---|---|
| 72 | 2 | TS 502 | B20 5011/14 | Cono diffusore | 32 |
| 73 | 2 | TS 257 | B20 5011A/68 | Centratore | 3,50 |
| 23 | 2 | TS 510 | B20 5011A/17 | Getto principale | 1,35 |
| 20 | 2 | TS 865 | B20 5011A/6 | Getti del minimo | 0,55 |
| 66 | 1 | TS 507 | B20 5011A/93 | Getto pompa | 0,65 |
| 71 | 1 | 1338 | B20 5011A/20 | Getto avviamento | 1,60 |
| 63 | 2 | TS 534a | B20 5011A/32 | Tubetto di emulsione | F 5 |
| 65 | 2 | 1311 | B20 5011A/19 | Getto aria di freno | 1,60 |
| 69 | 1 | 512 | B20 5011/85 | Getto aria avviamento | 5,00 |
| 15 | 1 | 1381a | B20 5011A/83 | Registro avviamento | F 2 |
| 12 | 1 | 1660a | B20 5011/9 | Valvola a spillo | 1,75 |

| Figura | Q. | DENOMINAZIONE | Numero di ordinazione WEBER | | Figura | Q. | DENOMINAZIONE | Numero di ordinazione WEBER | |
|---|---|---|---|---|---|---|---|---|---|
| 1 | 1 | Presa aria attacco filtro | TS 894 | B20 5011A/69 | 22 | 2 | Guarnizione per portagetto princip. | 2546 | B20 5011A/97 |
| 2 | 6 | Vite fissaggio coperchio carburatore | 665 | B20 5011B/41 | 23 | 2 | Getto principale | TS 510 | B20 5011A/17 |
| 3 | 6 | Rosetta elastica per vite di fissaggio | 41 | B20 5011B/28 | 24 | 2 | Portagetto principale | 2545 | B20 5011A/308 |
| 3A | 4 | Rosetta elastica per dado di fissaggio | 41 | B20 5011B/28 | 25 | 2 | Vite fissaggio centratore di miscela | TS 674 | B20 5011A/38 |
| 4 | 1 | Guarnizione per presa aria | A 1576 | B20 5011A/74 | 25A | 2 | Vite fissaggio cono diffusore | TS 674 | B20 5011A/33 |
| 5 | 1 | Guarnizione per corpo filtro | 236 | B20 5011A/53 | 26 | 2 | Vite ispezione fori accompagnamento | TS 625 | B20 5011A/84 |
| 6 | 1 | Reticella filtrante | 2494 | B20 5011A/44 | 27 | 1 | Vite registro andatura minimo | TS 548 | B20 5011A/34 |
| 7 | 1 | Guarnizione per corpo filtro | 235 | B20 5011A/95 | 28 | 1 | Corpo carburatore completo comprendente: | 2629a | B20 5011B/125 |
| 8 | 1 | Tappo raccordo porta filtro | 2612 | B20 5011A/43 | 29 | 1 | — Boccola di rasamento per alb. sinistro | 3565 | B20 5011B/149 |
| 9 | 1 | Guarnizione per coperchio carburatore | 2436 | B20 5011A/519 | 30 | 1 | — Spina conica fissaggio boccola | 2999 | B20 5011B/109 |
| 10 | 1 | Guarnizione per valvola a spillo | 1975 | B20 5011A/23 | 31 | 1 | — Molla per ritorno alberino sinistro | 3351 | B20 5011B/54s |
| 11 | 1 | Asta comando stantuffo pompa | 3558 | B20 5011B/90 | 32 | 1 | — Alberino principale sinistro completo | TS 539a | B20 5011A/1s |
| 12 | 1 | Valvola a spillo completa | 1660a | B20 5511/9 | 33 | 2 | — Valvola a farfalla | TS 555 | B20 5011A/15 |
| 13 | 1 | Piastrina ritegno molla stantuffo | 2903 | B20 5011B/105 | 34 | 4 | — Vite fissaggio valvola a farfalla | 767 | B20 5011/36 |
| 14 | 1 | Molla per stantuffo pompa | 2880 | B20 5011B/99 | 35 | 1 | — Alberino principale destro completo | TS 536a | B20 5011A/1d |
| 15 | 1 | Registro avviamento completo | 1381a | B20 5011A/83 | 36 | 1 | — Molla ritorno alberino destro | 2911 | B20 5011B/54d |
| 16 | 1 | Galleggiante completo | 4108a | B20 5011B/16 | 37 | 1 | Rosetta di sicurezza a doppia linguetta per alberino princip. destro | 2464 | B20 5011B/104 |
| 17 | 1 | Perno fulcro galleggiante | 1829 | B20 5011A/45 | | | | | |
| 18 | 1 | Valvola di aspirazione pompa completa | 937a | B20 5011A/110 | 37A | 1 | Rosetta di sicurezza a doppia linguetta fissaggio leva comando valvole a farfalla | 2464 | B20 5011B/104 |
| 19 | 1 | Stantuffo pompa completo | 2877a | B20 5011B/102 | | | | | |
| 20 | 2 | Getto del minimo | TS 865 | B20 5011A/6 | | | | | |
| 21 | 2 | Portagetto del minimo | TS 864 | B20 5011A/77 | | | | | |
| 38 | 1 | Dado fissaggio leva comando pompa | 865 | B20 5011A/13 | 55 | 1 | Vite per morsetto | 1144 | B20 5011A/114 |
| 38A | 1 | Dado fissaggio leva comando valvole a farfalla | 865 | B20 5011A/13 | 56 | 2 | Vite fissaggio coperchio valvola avv. | 336 | B20 5011A/111 |
| 39 | 1 | Leva comando pompa completa | TS 541a | B20 5011A/98 | 57 | 1 | Copiglia per perno valvola avviam. | 133 | B20 5011A/92 |
| 40 | 1 | Boccola distanziale per alber. destro | 3564 | B20 5011B/106 | 58 | 1 | Perno per leva valvola avviamento | 1585 | B20 5011A/101 |
| 41 | 1 | Settore dentato registrabile | TS 500 | B20 5011A/107 | 59 | 1 | Coperchio per valvola avviamento | TS 501 | B20 5011A/80 |
| 42 | 1 | Boccola registro settore dentato | 1350 | B20 5011A/91 | 60 | 1 | Molla per valvola avviamento | 939 | B20 5011A/76 |
| 43 | 2 | Molla per vite registro miscela minimo | 38 | B20 5011/25 | 61 | 1 | Valvola avviamento | 936 | B20 5011A/70 |
| | | | | | 62 | 1 | Sfera per alte velocità | 756 | B20 5011B/108 |
| 43A | 1 | Molla per vite registro andatura minimo | 38 | B20 5011/25 | 63 | 2 | Tubetto di emulsione | TS 534a | B20 5011A/32 |
| | | | | | 64 | 1 | Premisfera per valvola alte velocità | TS 1033 | B20 5011B/147 |
| 44 | 1 | Distanziale per vite registro andatura minimo | TS 546 | B20 5011A/94 | 65 | 2 | Getto aria di freno | 1311 | B20 5011A/19 |
| 45 | 2 | Vite registro miscela minimo | 864 | B20 5011B/40 | 66 | 1 | Getto pompa | TS 507 | B20 5011A/93 |
| 46 | 1 | Leva comando valvola avviamento completa comprendente: | 3554a | B20 5011B/146 | 67 | 1 | Vite fissaggio getto pompa | TS 621 | B20 5011A/112 |
| | | | | | 68 | 2 | Guarnizione per getto pompa e vite di fissaggio | 1520 | B20 5011A/96 |
| 47 | 1 | — Dado per vite fissaggio filo com. avviamento | 364 | B20 5011/73 | 69 | 1 | Getto aria avviamento | 512 | B20 5011/85 |
| 48 | 1 | — Leva comando valvola avviamento | TS 511 | B20 5011A/10 | 70 | 1 | Guarnizione per getto avviamento | 1339 | B20 5011A/2 |
| 49 | 1 | — Boccola per vite fissaggio filo comando avviamento | 375 | B20 5011/67 | 71 | 1 | Getto avviamento | 1338 | B20 5011A/20 |
| | | | | | 72 | 2 | Cono diffusore | TS 502 | B20 5011A/14 |
| 50 | 1 | — Vite fissaggio filo comando avviamento | 374 | B20 5011B/39 | 73 | 2 | Centratore di miscela | TS 257 | B20 5011A/68 |
| | | | | | 74 | 1 | Coperchio carburatore completo comprendente: | 2634a | B20 5011A/88 |
| 51 | 1 | Coperchio scatola settori dentati | 2617 | B20 5011A/48 | 75 | 1 | — Coperchio carburatore | 3025a | — |
| 52 | 1 | Leva comando valvole a farfalla completa | LP 197a | B20 5011A/2 | 76 | 4 | — Vite prigioniera fissaggio presa aria | G 228 | B20 5011A/150 |
| 53 | 1 | Morsetto fissaggio sett. dentato reg. | 1335 | B20 5011A/100 | 77 | 4 | Dado per vite prigioniera fiss. presa aria | 340 | B20 5011/72 |
| 54 | 2 | Vite fissaggio cop. scat. settori dent. | 2288 | B20 5011A/33 | | | | | |

## CARBURATOR WEBER 40 DCL/6

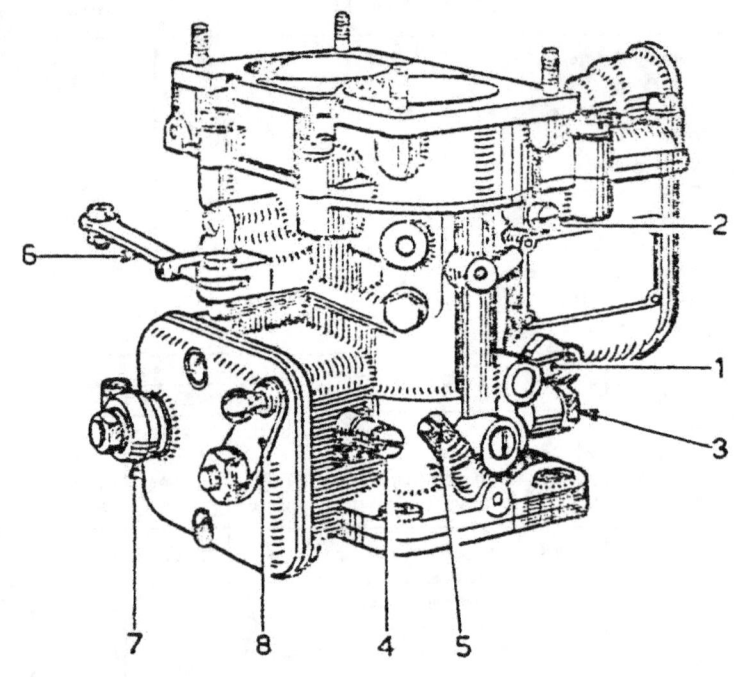

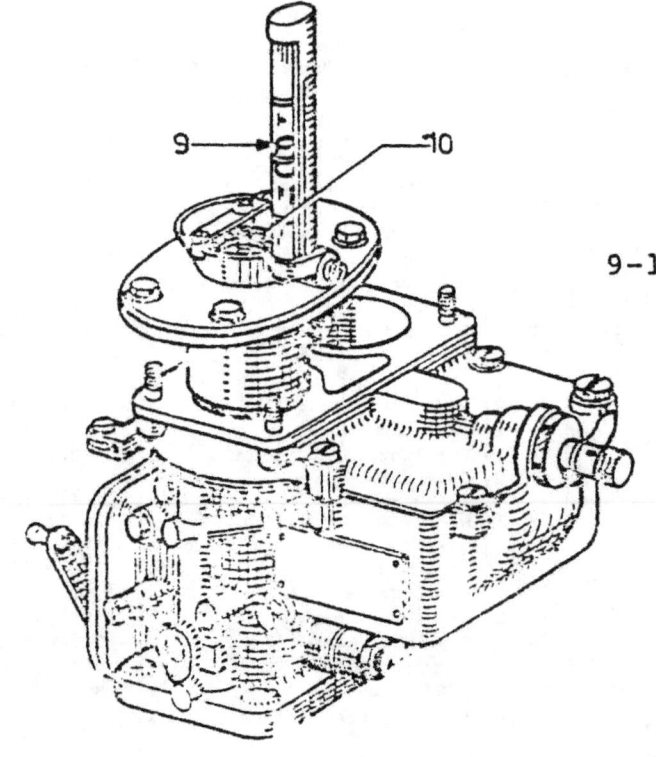

### LEGEND

1. MAIN JET
2. IDLE JET
3. ACCELLERATOR PUMP
4. MINIMUM THROTTLE SETTING
5. IDLE MIXTURE
6. CHOKE
7. SECOND BUTTERFLY ADJ.
8. THROTTLE LINKAGE
9-10. UNI-SYN ADJ. TOOL

 WEBER CARBURETORS

Types 36 - 40 - 42 DCF/DCL/DCZ

36/40/42 DCF DCL DCZ

| | |
|---|---|
| Type | twin choke downdraft |
| Intake pipe diameter in mm. | 36 - 40 - 42 |
| Starting device | with E.I. (E = Summer - I = Winter) control |
| Accelerating pump | metal piston |
| Extra power device | by valve |
| Metal | DCF: anticorodal aluminium (production ceased)<br>DCL: pressure cast anticorodal aluminium<br>DCZ: pressure cast zamac alloy |

Some of the most popular installations: **Alfa Romeo 1900 - Bugatti 101 Ferrari 166/212/250/340 - Fiat 8V - Lancia Aurelia B20/B22 Pegaso Z 102**

### INTRODUCTION

The double choke downdraft carburetors of the « DCF - DCL - DCZ » type are obtainable with diameters at the height of the throttle valves of 36, 40 and 42 mm. thus permitting their use over a great range of engines.
On these carburetors the device for governing the fuel mixture consists of two throttles mounted on two parallel shafts. These valves are kept in perfect relation by means of two geared segments mounted on the ends of the shafts and they open and close in counter rotation assuring a perfectly equal fuel distribution in either intake pipe.
The carburetors of the « DCF - DCL - DCZ » type are provided with an accelerating pump and starting device; moreover on request, they can be supplied with the full power device fitted. The main intake pipes of this type of carburetor work independently one from the other, since each of them constitutes a complete single choke carburetor.

### DESCRIPTION

The cross section in **Figure 1**, shows how the air arrives from the top, passes through the auxiliary Venturi (2) where it mixes with the fuel coming out from the discharge tubes (3) and then, through the chokes (30) it is carried to the engine cylinders according to the opening of the throttle valves (28). From the fuel line connected with the carburetor by means of a suitable fitting, the fuel flows through the needle valve (11) into the float bowl (16) where the float (10), hinged to the pivot (13), controls the needle opening (12) and maintains the fuel level constant.
From the float bowl the fuel controlled by the calibrated main jets (22) arrives at the emulsioning tubes (4) by means of the pipes (23) from which mixed with the air arriving from the calibrated air adjusting screws (5) through the emulsioning tubes and discharge tubes (3) it reaches the carburation zone constituted by the auxiliary Venturis (2) and by the chokes (30).
The purpose of the auxiliary Venturis is to increase the vacuum around the discharge tubes (3) and to carry the emulsified fuel to the center of the chokes (30) at their narrowest diameter so as to render the mixture more homogeneous with the advantage of a better distribution to the cylinders.
For idling speed operation of the engine the fuel, by means of suitable pipes, is carried from the emulsioning tubes (4) to the calibrated idling jets (6) from which, emulsified with the air deriving from the calibrated holes (7) through the tubes (29) and the idling feeding holes (27) adjustable by means of conical screws, it arrives at the carburetor throttles chamber below the throttles where it mixes itself with the air which is sucked in by the engine vacuum through the small openings existing between the throttles chamber walls and the throttles when in idling position. From the tubes (29) the mixture arrives at the carburetor throttles chamber through the progression holes (26) situated in relation to the throttles and having the purpose of permitting smooth increase of the engine speed when starting from idle, when the throttles are opened.
The accelerating pump permits a regular increase of engine speed even when the throttles are suddenly opened.
In the carburetors of the « DCF-DCL-DCZ » type the accelerating pump is a metal piston (15) activated by the pump control shaft (8) through the lever with small roller (25) fixed to the shaft bearing the throttles lever.
When closing the throttles, the lever (25) by means of the shaft (8) lifts up the piston (15); the fuel is then drawn from the float bowl into the pump cylinder through the intake valve (19). Opening the throttles, the shaft (8) remains free and the piston (15) is pushed towards the bottom by the spring (9); by means of the tube (17) the fuel is forced through the ball delivery valve (31) to the pump jets body (1) from which it is injected into the carburetor main intake pipes by means of suitable calibrated tubes.
In order to vary the fuel quantity discharged by the accelerating pump, the carburetors of the « DCL-DCZ » type are provided with a pump exhaust screw (18); for the DCF type the exhaust is obtained by means of a suitable hole made in the pump piston (15).
In the carburetors of the « DCF-DCL-DCZ » type the ball check of the delivery valve of the accelerating pump may be substituted by a needle valve (14).
In case of special need, that is when each carburetor intake pipe feeds three cylinders or more, the carburetors of the « DCF-DCL-DCZ » type may be supplied complete with the full power device constituted by the valve (24) and calibrated jets (20). With throttles completely open the piston (15) pushed towards the bottom by the spring (9) opens the full power valve (24) allowing the fuel calibrated by the jets (20) to pass from the float bowl to the opening of the emulsioning tubes (4) through the intake valve (19) and the tubes (21) thus increasing the mixture strength drawn in by the engine through the auxiliary Venturis (2).

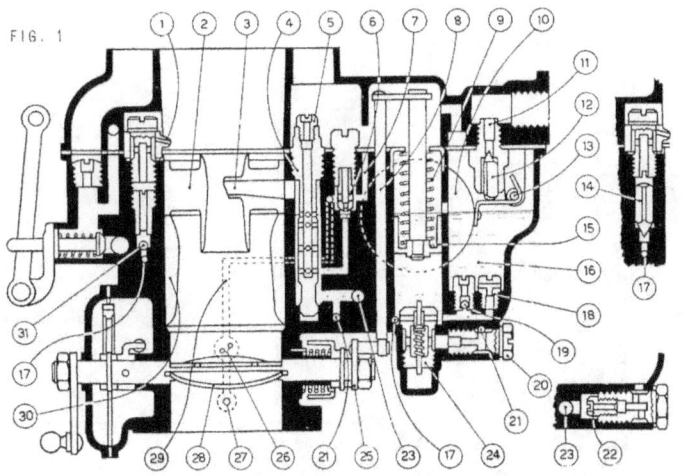

FIG. 1

### DESCRIPTIVE CROSS - SECTION

1 - Pump jets body
2 - Auxiliary Venturi
3 - Discharge tube
4 - Emulsioning tube
5 - Air adjusting screw
6 - Idling Jet
7 - Idling air hole
8 - Pump control shaft
9 - Pumping prolongation spring
10 - Float
11 - Needle valve seat
12 - Needle for valve
13 - Float fulcrum pivot
14 - Needle delivery valve of pump
15 - Pump piston
16 - Float bowl
17 - Pump tube
18 - Pump exhaust screw
19 - Pump intake valve
20 - Full power jet
21 - Full power tube
22 - Main jet
23 - Jet-Emulsion tube pipe
24 - Full power valve
25 - Pump control lever
26 - Progression holes
27 - Idling hole to the intake pipe
28 - Throttle
29 - Idling mixture tube
30 - Choke
31 - Ball delivery valve

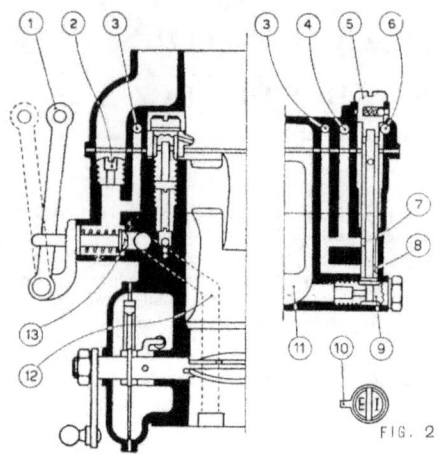

FIG. 2

## DESCRIPTIVE CROSS SECTION

1 - Starting valve control lever
2 - Starting air screw
3 - Starting mixture hole
4 - Starting air hole
5 - Starting control
6 - Air hole for starting control
7 - Hole for Summer mixture
8 - Hole for Winter mixture
9 - Starting jet
10 - Reference mark of the control on the carburetor cover
11 - Float bowl
12 - Starting mixture tube
13 - Starting valve

### STARTING DEVICE - Figure 2

The starting device allows a quick start when engine is cold. It is controlled from normal driving position by pulling a suitable knob on the instrument panel and should be released as soon as the engine reaches a sufficient temperature for regular running.
The fuel flowing from the constant level float bowl (11) through the calibrated jet (9) arrives at the housing tube (5) of the starting device. With the throttles in idling position the conical valve (13) is opened by the lever (1) the suction due to the engine under starter operation consents that the fuel, after a primary emulsification with the air coming through the holes (4) and (6), reaches the conical valve port (13) through the tube (3). This mixture is then further emulsified by air drawn through the calibrated screw (2) and is carried, by means of the tube (12), to the carburetor main pipes below the throttles.
For correct operation of the device it is necessary that the letter engraved upon the control (5), giving the weather conditions (E = Summer - I = Winter) should be in index with the reference finger (10) on the carburetor cover; the mixture formed by the fuel arriving from the jet (9), and the air arriving from the hole (6) is in this way fed by the calibrated hole (7) - Summer position - or by the calibrated hole (8) - Winter position - so that the device may supply the proper mixture for quick starting of the engine.

In order to allow the user to visualize the external pieces forming the carburetors of the «DCF - DCL - DCZ» type described in Figures 1 and 2, a sideview of the carburetor is shown in Figure 3 from which is seen that said pieces are easily accessible and demountable.
The carburetor being symmetrical in respect to a plane passing between the main pipes, said view denotes also the other side of the carburetor with the exception of the starting jet that, as already mentioned, is unique.

1 - Fuel filter casing
2 - Starting mixture control
3 - Idling jet
4 - Auxiliary Venturi securing screw
5 - Choke securing screw
6 - Idling mixture adjusting screw
7 - Idling speed adjusting screw
8 - Idling mixture adjusting screw
9 - Inspection screw for progression holes
10 - Main jet
11 - Full power jets
12 - Starting jet
13 - Bolt securing fuel filter casing

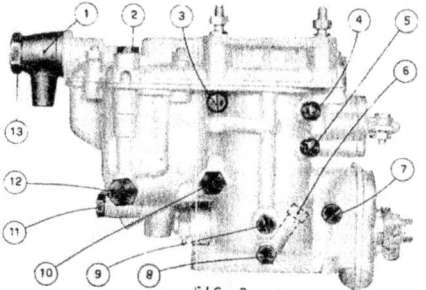

FIG. 3

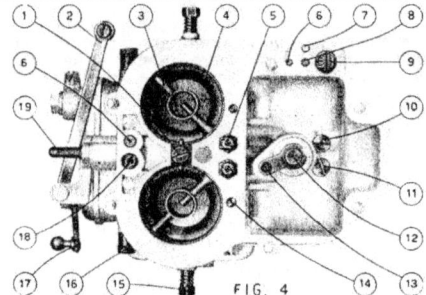

FIG. 4

In order to indicate the position of the internal pieces of the carburetors of «DCF - DCL - DCZ» type Figure 4 shows the plan view of said carburetor without the cover.

1 - Pump jets body
2 - Starting control lever
3 - Auxiliary Venturi
4 - Chokes
5 - Emulsioning tubes complete with air adjusting screws
6 - Starting mixture tube
7 - Starting emulsion air hole
8 - Starting emulsioning tube
9 - Starting control
10 - Pump intake valve
11 - Pump exhaust screw (for DCL-DCZ only)
12 - Accelerating pump
13 - Pump control shaft
14 - Bushings for idling air
15 - Idling mixture adjusting screws
16 - Idling air adjusting screws
17 - Throttle control lever
18 - Starting emulsion air screw
19 - Starting valve

## TUNING FOR IDLING

In the carburetor of «DCF - DCL - DCZ» type (Figure 5) the idle adjustment device consists of the idling speed adjusting screws (1) and the mixture adjusting screws (2). The screws (1) control the amount of throttles opening syncronized by means of the geared segments; the screws (2) with the conical end maintain the proper air-fuel ratio for smooth engine operation by controlling the quantity of the mixture from the idling jets and its mixture with the air drawn in by the engine. The screws (2) can be arranged as indicated in the sketch (6) of Figure 3.
Tuning for the proper idle must be carried out with engine warmed setting first the minimum opening of the throttle valves by means of the screws (1) adjusting it to a position to prevent the engine from stalling under all conditions. Then turn the screw (2) to obtain the best mixture strength for the fastest, stable and smoothest running at that throttle valves position. Finally the throttle valves opening can then be reduced further to obtain the most suitable idling speed.
In Figure 5 is also shown the boss (3) for the eventual connection of the automatic spark advance. For fitting on engines having said device, the carburetors of «DCF - DCL - DCZ» type can be supplied, at request, completed with said connecting piece.

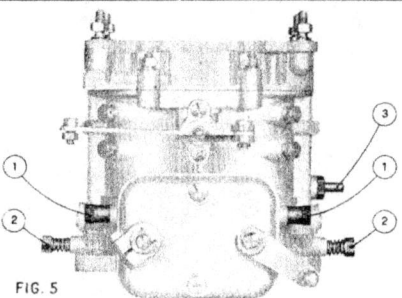

FIG. 5

## OVERALL DIMENSIONS in mm.

## CARBURETOR FIXING FLANGES

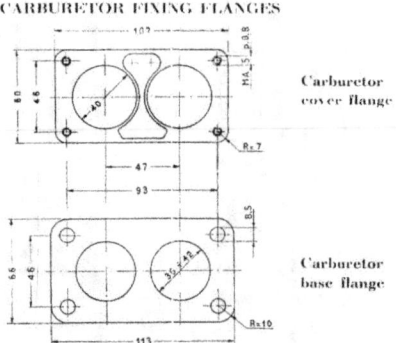

Carburetor cover flange

Carburetor base flange

Registered Office in Milan - Works: BOLOGNA, Via Timavo 33 - (ITALY)
Cable Adress: WEBER - BOLOGNA - Telephone 64 5 73

# CARBURATORI WEBER
## Tipo 36 DCS

**Applicazione**
FERRARI 250 GT/E. 62

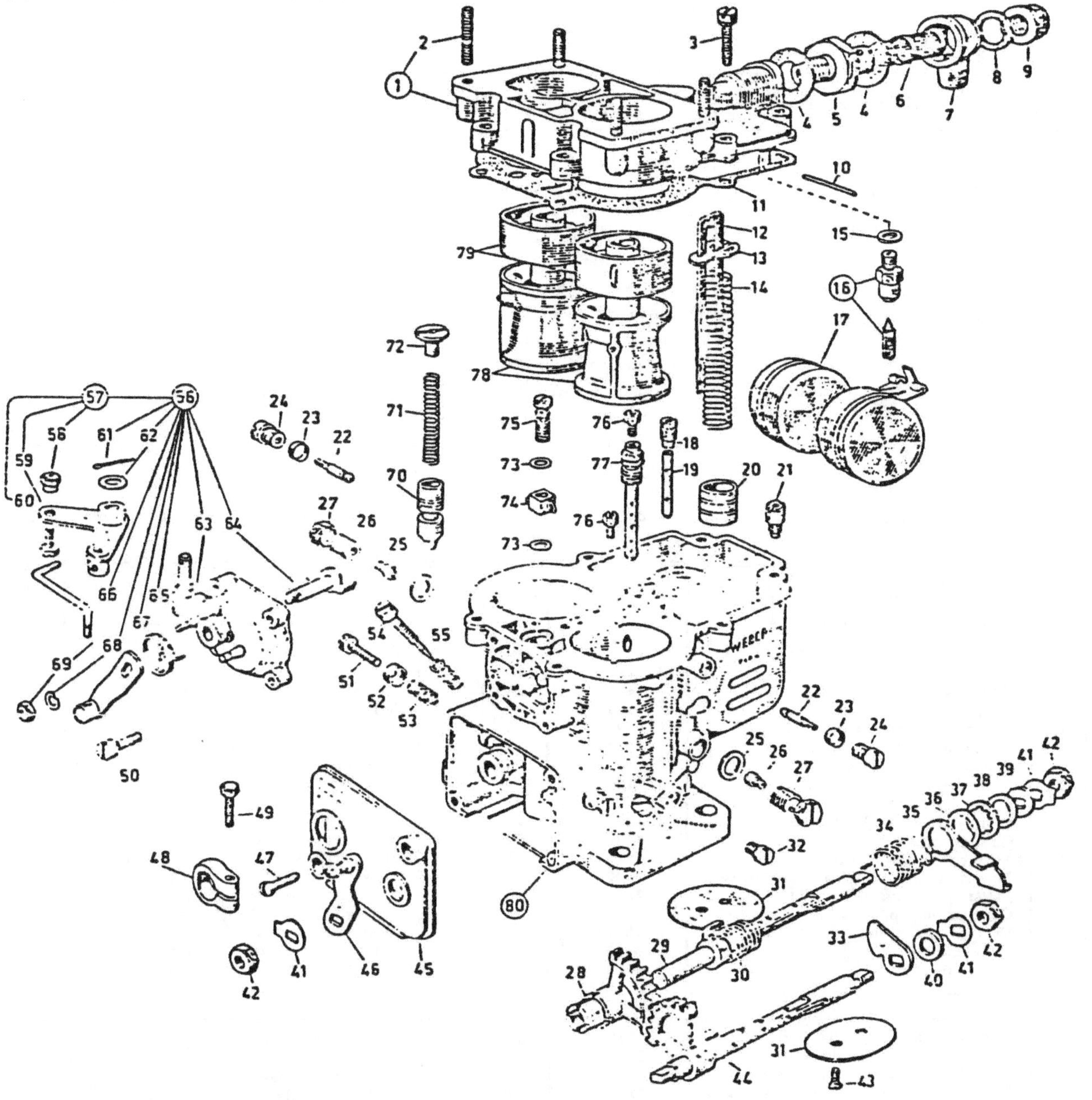

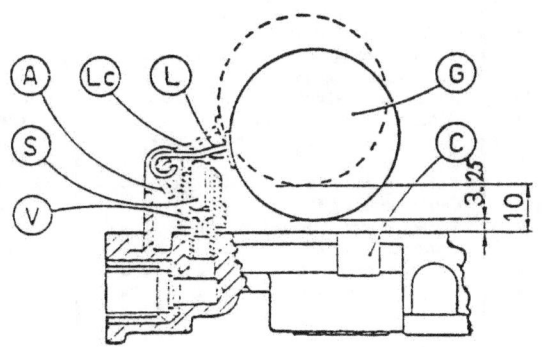

Nelle ordinazioni indicare: numero di matricola del particolare richiesto, la eventuale taratura e il numero e tipo del carburatore.

Per effettuare la livellatura del galleggiante è necessario attenersi alle seguenti norme di carattere generale:

- accertarsi che la valvola a spillo (V) sia bene avvitata nel suo alloggiamento;
- con coperchio carburatore (C) capovolto, i due semigalleggianti devono distare dal piano del coperchio stesso mm. 3.25
- a livellatura effettuata, controllare che la corsa del galleggiante (G) sia di mm. 7 modificando eventualmente la posizione dell'appendice (A).
- Qualora il galleggiante (G) non fosse giustamente impostato, modificare la posizione delle linguette (L) del galleggiante stesso fino a raggiungere la quota richiesta, avendo cura che la linguetta (Lc) sia perpendicolare all'asse dello spillo (S) e che non presenti, sul piano di contatto, incaccature che possano influire sul libero scorrimento dello spillo stesso.

# CARBURATORI WEBER
## Tipo 36 DCS

**Applicazione**
**FERRARI 250 GT/E. 62**

## REGOLAZIONE

| Figura | Quantità | Numero di ordinazione | DENOMINAZIONE | Taratura in mm. |
|---|---|---|---|---|
| 78 | 2 | 3504a | Cono diffusore . . . . . . . . . . . . . . . . . . . | 27 |
| 79 | 2 | 3495a | Centratore . . . . . . . . . . . . . . . . . . . . | 3 |
| 26 | 2 | TS 510 | Getto principale . . . . . . . . . . . . . . . . . | 1,40 |
| 22 | 2 | TS 865 | Getto minimo . . . . . . . . . . . . . . . . . . | 0,55 |
| 74 | 1 | 4823 | Getto pompa . . . . . . . . . . . . . . . . . . . | 0,60 |
| 19 | 1 | 4844 | Getto avviamento . . . . . . . . . . . . . . . . . | F1/0,60 |
| 77 | 2 | TS 534a | Tubetto emulsionatore . . . . . . . . . . . . . . | F 8 |
| 76 | 2 | 1311 | Getto aria di freno . . . . . . . . . . . . . . . | 2,40 |
| 18 | 1 | 4843 | Getto aria avviamento . . . . . . . . . . . . . . | 1,50 |
| 16 | 1 | 1660a | Valvola a spillo . . . . . . . . . . . . . . . . . | 1,75 |
| 21 | 1 | 2943a | Vite di aspirazione con foro di scarico . . . . . . | « chiuso » |
| 17 | 1 | 6607a | Galleggiante . . . . . . . . . . . . . . . . (peso) | gr. 18 |
| — | — | — | Livellatura galleggiante . . . . . . . . . . . . . | 3 (**) |

La S.p.A. E. WEBER non risponde di anomalie di funzionamento dovute ad arbitrarie modifiche apportate alla regolazione indicata nel presente Catalogo.

| Figura | Q. | DENOMINAZIONE | Numero di ordinazione | Figura | Q. | DENOMINAZIONE | Numero di ordinazione |
|---|---|---|---|---|---|---|---|
| 1 | 1 | Coperchio carburatore completo di: | 6994a | 19 | 1 | Getto avviamento . . . . . . . . | 4844 |
| 2 | 4 | — Vite prigioniera . . . . . . . . | G 228 | 20 | 1 | Stantuffo pompa . . . . . . . . | 2877a |
| 3 | 6 | Vite fissaggio coperchio carburatore | 665 | 21 | 1 | Valvola aspirazione pompa . . . | 2743a |
| 4 | 2 | Guarnizione . . . . . . . . . . . | 235 | 22 | 2 | Getto del minimo . . . . . . . . | TS 865 |
| 5 | 1 | Raccordo porta filtro . . . . . . | TS 597 | 23 | 2 | Anello di tenuta portagetto minimo | 3456 |
| 6 | 1 | Reticella filtrante . . . . . . . . | 241a | 24 | 2 | Portagetto minimo . . . . . . . | 3470 |
| 7 | 1 | Corpo filtro . . . . . . . . . . | F 23 | 25 | 2 | Guarnizione . . . . . . . . . . | 2546 |
| 8 | 1 | Guarnizione . . . . . . . . . . . | 236 | 26 | 2 | Getto principale . . . . . . . . | TS 510 |
| 9 | 1 | Dado per raccordo . . . . . . . | TS 598 | 27 | 2 | Portagetto principale . . . . . . | 2545 |
| 10 | 1 | Perno fulcro galleggiante . . . . | 1829 | 28 | 1 | Settore dentato registrabile . . . | 7020 |
| 11 | 1 | Guarnizione per coper. carburatore | 4790 | 29 | 1 | Alberino sinistro . . . . . . . . | 7001a |
| 12 | 1 | Asta comando pompa . . . . . . | 6998 | 30 | 1 | Molla ritorno alberino sinistro . . | 7166 |
| 13 | 1 | Piastrina ritegno molla stantuffo . | 3500 | 31 | 2 | Valvola a farfalla . . . . . . . . | 915 |
| 14 | 1 | Molla per stantuffo pompa . . . | 2580 | 32 | 2 | Vite ispezione fori di progressione | 6999 |
| 15 | 1 | Guarnizione per valvola a spillo . | 1975 | 33 | 1 | Camma comando pompe . . . . | 7000 |
| 16 | 1 | Valvola a spillo . . . . . . . . . | 1660a | 34 | 1 | Molla ritorno leva allentata . . . | 7242 |
| 17 | 1 | Galleggiante . . . . . . . . . . | 6607a | 35 | 1 | Leva allentata comando pompa . | 6512 |
| 18 | 1 | Getto aria di freno avviamento . | 4843 | 36 | 1 | Rosetta per anello elastico . . . | 6584 |
| 37 | 1 | Anello elastico dentellato . . . . | 6525 | 59 | 1 | — Leva per boccola . . . . . . | 7017a |
| 38 | 1 | Rosetta per anello elastico . . . | 6527 | 60 | 1 | — Vite fissaggio filo . . . . . . | 4102 |
| 39 | 1 | Rosetta distanziale per alberino sin. | 4 | 61 | 1 | — Copiglia spaccata . . . . . . | 133 |
| 40 | 1 | Boccola per alberino destro . . . | 7171 | 62 | 1 | — Rosetta piana . . . . . . . . | 7009 |
| 41 | 3 | Rosetta di sicurezza . . . . . . | 6371 | 63 | 1 | — Coperchio con perno . . . . | 7010a |
| 42 | 3 | Dado fissaggio per alberini . . . | 865 | 64 | 1 | — Alberino comando valvola avv. | 4935a |
| 43 | 4 | Vite fissaggio valvola a farfalla . | 767 | 65 | 1 | — Tirantino . . . . . . . . . . | 7008 |
| 44 | 1 | Alberino destro . . . . . . . . . | 7003a | 66 | 1 | — Molla ritorno leva . . . . . . | 3473 |
| 45 | 1 | Coperchio scatola settore dentati . | 4821 | 67 | 1 | — Leva di comando . . . . . . | 7013a |
| 46 | 1 | Leva comando farfalla . . . . . | 7005a | 68 | 1 | — Rosetta elastica . . . . . . . | 6983 |
| 47 | 2 | Vite fissaggio coper. scatola settori | 2288 | 69 | 1 | — Dado esagono . . . . . . . . | 335 |
| 48 | 1 | Morsetto fissaggio settore dentato | 4837 | 70 | 1 | Valvola avviamento . . . . . . . | 5957 |
| 49 | 1 | Vite fissaggio morsetto . . . . . | A 1408 | 71 | 1 | Molla per valvola avviamento . . | 5248 |
| 50 | 4 | Vite fissaggio comando avviamento | 1153 | 72 | 1 | Ritegno e guida molla . . . . . | 4841 |
| 51 | 2 | Vite registro farfalla . . . . . . | 3477 | 73 | 2 | Guarnizione per getto pompa . . | 1520 |
| 52 | 2 | Distanziale per vite registro farfalla | 7133 | 74 | 1 | Getto pompa . . . . . . . . . . | 4823 * |
| 53 | 2 | Molla per vite registro farfalla . . | 1173 | 75 | 1 | Valvola mandata pompa . . . . | TS 991a |
| 54 | 2 | Vite registro miscela minimo . . | 864 | 76 | 2 | Getto aria di freno . . . . . . . | 1311 |
| 55 | 2 | Molla per vite registro miscela min. | 38 | 77 | 2 | Tubetto emulsionatore . . . . . | TS 534a * |
| 56 | 1 | Comando avviamento completo di: | 7007a | 78 | 2 | Cono diffusore . . . . . . . . . | 3504a * |
| 57 | 1 | — — Leva comando tirantino completa dei particolari . . . . . . | 7015a | 79 | 2 | Centratore . . . . . . . . . . . | 3495a * |
| 58 | 1 | — Dado per vite . . . . . . . . | 4103 | 80 | 1 | Corpo carburatore . . . . . . . | Non a ricambio |

(*) Particolari tarati di regolazione.

# CARBURATORI WEBER

36 DCS

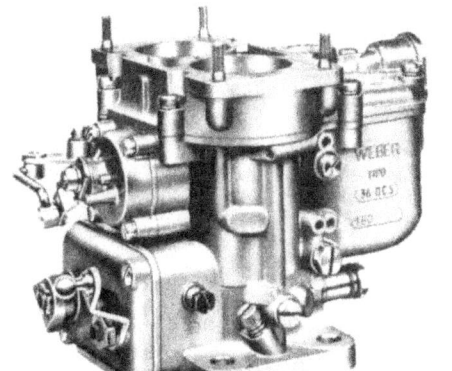

CARBURATORI Tipo | **36 DCS**
CARBURETORS Type |

Applicazione | **FERRARI**
Standard Equipment on | 250 GT/E. 62

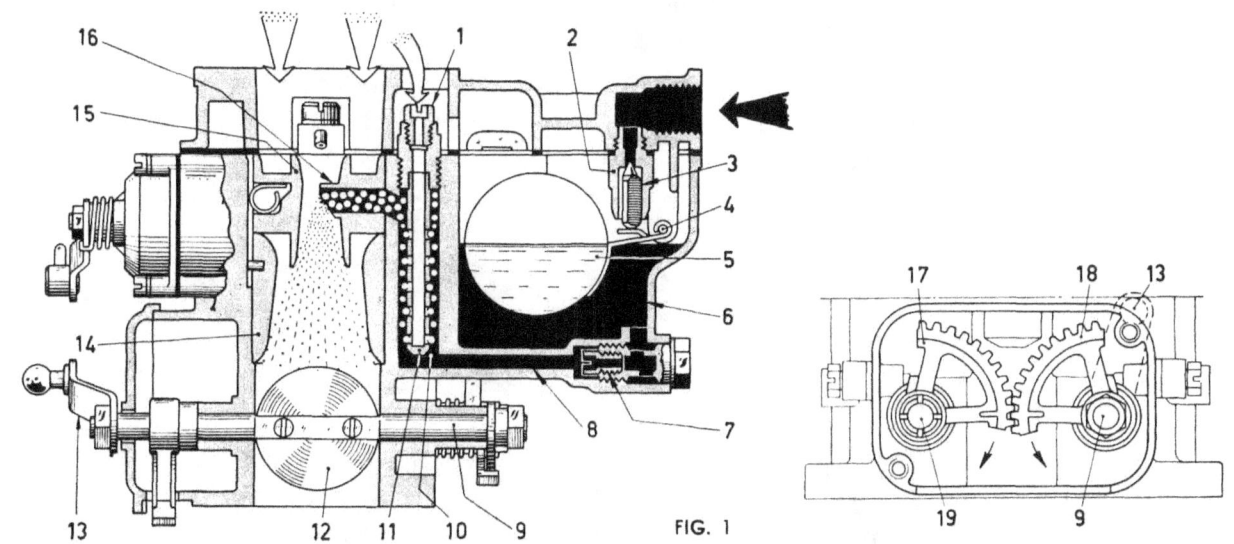

FIG. 1

### MARCIA NORMALE - Fig. 1

Il carburante, attraverso la valvola a spillo (2), passa alla vaschetta (6), dove il galleggiante (5), articolato nel perno fulcro (4), regola l'apertura dello spillo (3), per mantenere costante il livello del liquido. Dalla vaschetta (6), attraverso i getti principali (7) ed i canali (8), il carburante giunge ai pozzetti (10), ove miscelato con l'aria uscente dai fori dei tubetti emulsionatori (11) e proveniente dai getti aria di freno (1), giunge attraverso i tubetti spruzzatori (16), alla zona di carburazione, costituita dai centratori (15) e dai diffusori (14).

In fig. 1 è illustrato il dispositivo per l'apertura sincronizzata delle valvole a farfalla. Agendo sulla leva (13), le farfalle (12) vengono comandate in modo sincrono mediante i settori dentati (17) e (18), fissati sugli alberini (19) e (9) e si aprono una in senso contrario all'altra, garantendo così una perfetta simmetria di alimentazione ai condotti di ammissione.

### NORMAL RUNNING - Fig. 1

The fuel, through the needle valve (2) passes to the bowl (6) where the float (5), articulated in the trunnion (4), regulates the needle opening (3) in order to keep the level of the liquid constant. From the bowl (6), through the main jets (7) and ducts (8), the fuel reaches the wells (10) where, mixed with the air from the orifices of the emulsioning tubes (11) and coming from the air corrector jets (1), through the nozzles (16), it reaches the carburation area, consisting of the venturi (15) and the secondary venturi (14).

Fig. 1 shows the device for synchronized opening of the throttles. Acting on lever (13), the throttles (12) are syinchronously controlled by means of toothed sectors (17) and (18) fixed to spindles (9) and (19), and open in opposite directions, so making sure of perfectly even feeding to the inlet ducts.

30-12-1964 · PRINTED IN ITALY

CATALOGO GENERALE WEBER

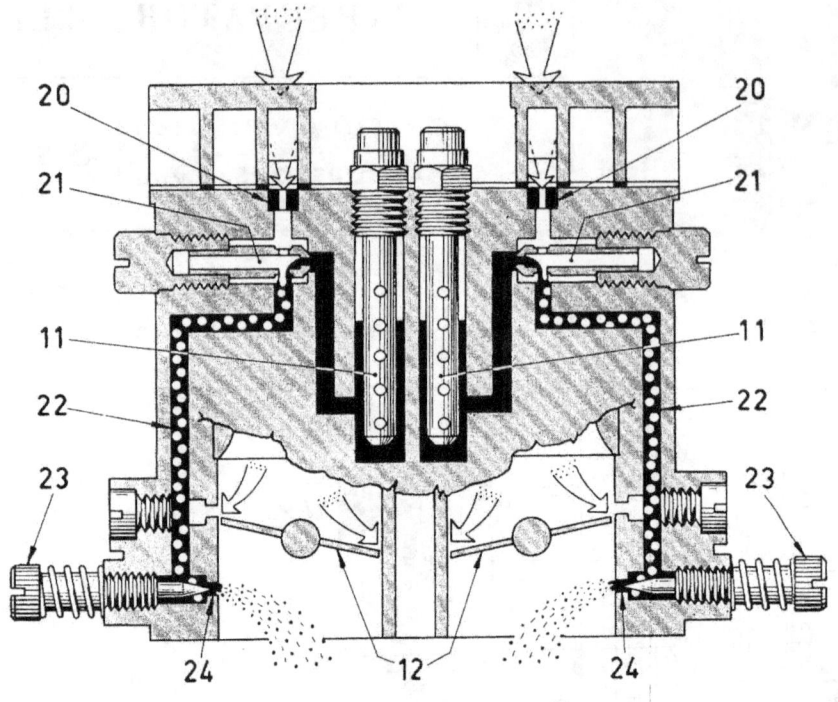

FIG. 2

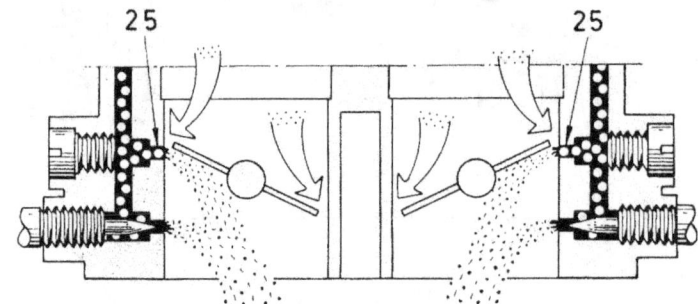

### MARCIA AL MINIMO E PROGRESSIONE - Fig. 2

Il carburante passa dai pozzetti, dei tubetti emulsionatori (11), ai getti del minimo (21). Emulsionato con l'aria proveniente dalle boccole calibrate (20), giunge attraverso i canali (22) ed i fori alimentazione minimo (24) registrabili mediante le viti (23) ai condotti del carburatore, a valle delle farfalle (12).

La miscela giunge ai condotti, anche dai fori di progressione (25), posti in corrispondenza delle farfalle, permettendo così un regolare aumento della velocità angolare del motore, a partire dal regime di minimo.

### IDLE SPEED AND PROGRESSION - Fig. 2

From the primary emulsioning tube wells (11) the fuel passes to the idle jets (21) from which, emulsioned with the air coming from the calibrated bush (20), through ducts (22) and the idle feed orifices (24), the last being adjustable by means of screws (23), it reaches the carburetor ducts downstream of the throttles (12).

The mixture also reaches the ducts from progression holes (25) placed on a level with the throttles, so allowing a regular increase in angular speed of the engine starting from idling speed.

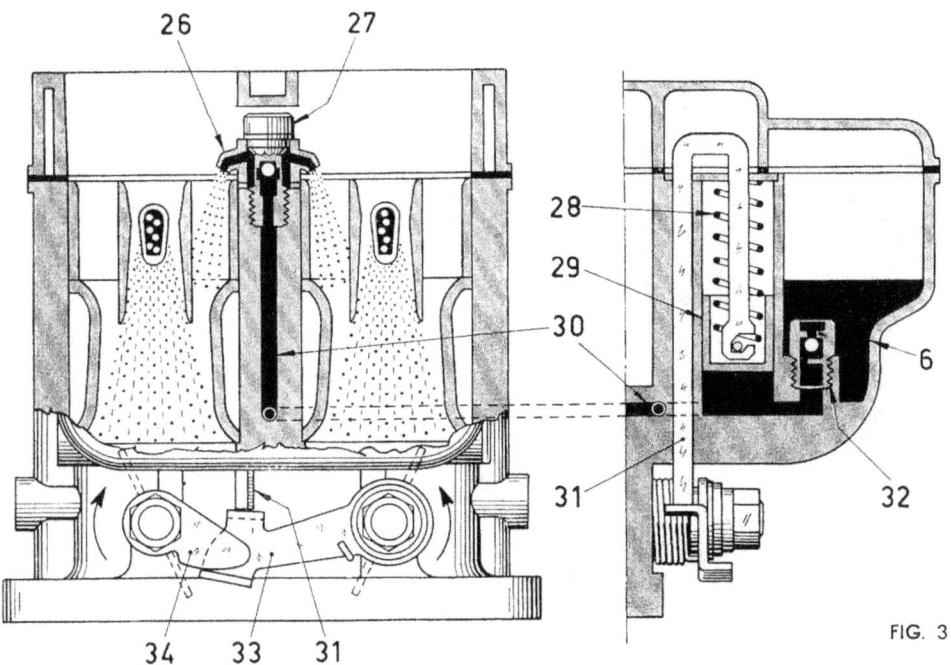

FIG. 3

**FUNZIONAMENTO IN ACCELERAZIONE** - Fig. 3

Chiudendo le farfalle, la leva (**33**), tramite l'asta (**31**) solleva lo stantuffo (**29**); il carburante viene aspirato dalla vaschetta (**6**) nel cilindro della pompa, attraverso la valvola di aspirazione (**32**).

Aprendo le farfalle, la leva (**34**) abbassa la leva (**33**), liberando l'asta (**31**). Lo stantuffo (**29**), sotto l'azione della molla (**28**), viene spinto verso il basso; mediante la conduttura (**30**) il carburante viene iniettato attraverso la valvola (**27**) ed i tubetti tarati del getto pompa (**26**) nei condotti del carburatore.

La valvola di aspirazione (**32**), può essere provvista di un foro laterale calibrato, che scarica in vaschetta l'eccesso di carburante.

**ACCELERATION** - Fig. 3

Closing the throttles, lever (**33**), through the rod (**31**), raises the plunger (**29**). The fuel is drawn from the bowl (**6**) into the cylinder of the pump through the inlet valve (**32**). By opening the throttles, lever (**34**) lowers lever (**33**), so freeing rod (**31**). The plunger (**29**), through the action of spring (**28**), is pushed downwards; along the ducts (**30**) the fuel is injected through valve (**27**) and the calibrated pipes of the jet pump (**26**) into the carburetor ducts.

The inlet valve (**32**) may be supplied with a lateral calibrated orifice which discharges any excess fuel into the well.

**MISURE D'INGOMBRO** in mm. — **OVERALL DIMENSIONS** in mm.

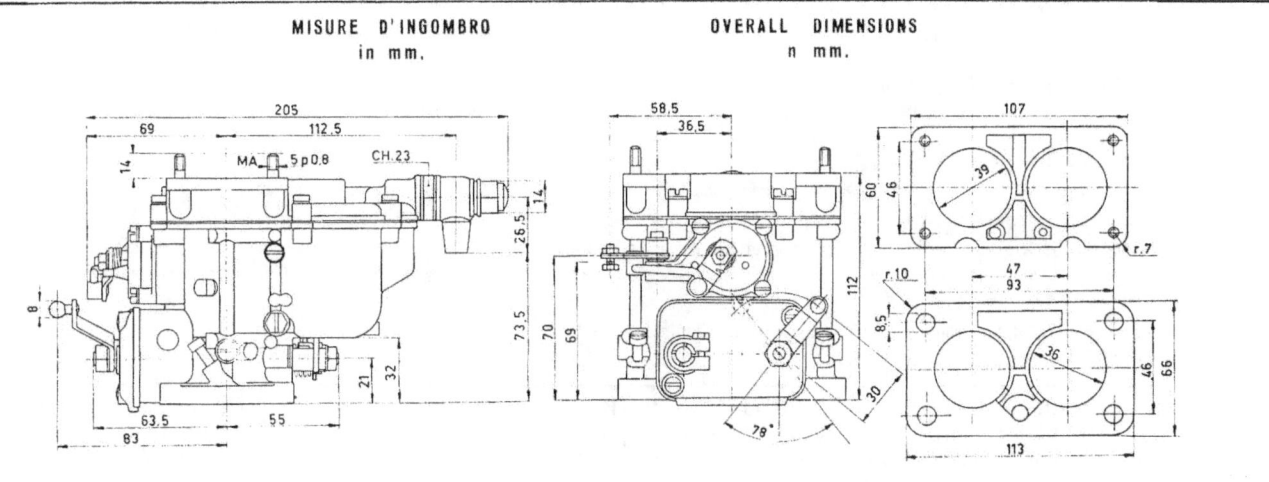

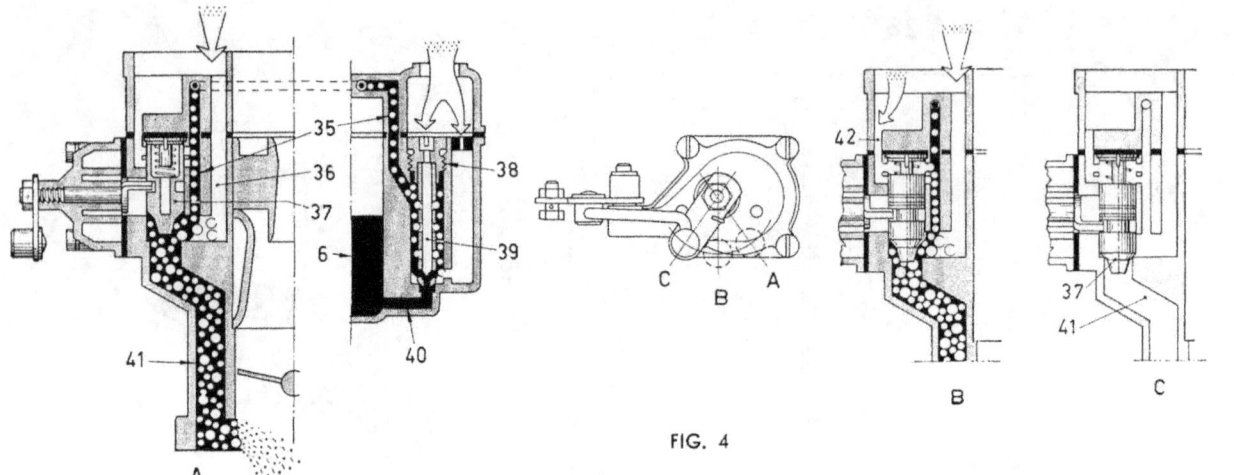

FIG. 4

## DISPOSITIVO DI AVVIAMENTO  Fig. 4

Il carburante, dalla vaschetta (6), passa al dispositivo di avviamento, attraverso il canale (40) ed il getto avviamento (39). Emulsionato con l'aria calibrata dal getto aria di freno (38) proveniente dalla presa d'aria del carburatore giunge al vano dello stantuffo (37), attraverso il canale (35), ove si miscela con l'aria proveniente dal canale (36). La miscela così formata, viene aspirata attraverso il canale (41), permettendo un pronto avviamento del motore. (Schema A)

Ad avviamento ottenuto, disinserire parzialmente il dispositivo di avviamento. (Schema B)

In queste condizioni, un ulteriore afflusso di aria proveniente dal canale (42), smagrisce il titolo della miscela erogata dal dispositivo di avviamento, permettendo un regolare funzionamento del motore a freddo.

Riscaldandosi però il motore, detta miscela risulta a titolo troppo ricco ed in quantità eccessiva; pertanto è necessario escludere progressivamente il dispositivo di avviamento, con l'aumentare della temperatura del motore.

Con il dispositivo di avviamento disinserito, lo stantuffo (37) chiude il canale (41), impedendo il richiamo di miscela. (Schema C)

## NORME DI IMPIEGO DEL DISPOSITIVO AVVIAMENTO

Per ottenere dal dispositivo avviamento, tutti i vantaggi che esso può fornire, si riassumono le norme di impiego, che è opportuno osservare.

## AVVIAMENTO DEL MOTORE

**Avviamento a freddo.** Inserire completamente il dispositivo di avviamento. Posizione «A». Ad avviamento ottenuto ridurne il grado d'inserzione.

**Avviamento a motore semicaldo.** In questo caso è sufficiente inserire parzialmente il dispositivo di avviamento. Posizione «B».

**Messa in efficienza del veicolo.** Durante il periodo di riscaldamento del motore, anche con veicolo in moto, disinserire progressivamente il dispositivo con manovre successive, in modo da avere sempre una erogazione di miscela supplementare strettamente necessaria per un regolare funzionamento del motore. Posizione «B».

**Marcia normale del veicolo.** Non appena il motore ha raggiunto una temperatura sufficiente per un regolare funzionamento, escludere il dispositivo di avviamento. Posizione «C».

## STARTING DEVICE - Fig. 4

The fuel passes from the bowl (6) to the starting device through duct (40) and the starting jet (39). Emulsioned with the air coming from the carburetor air intake (calibrated by the air corrector jet (38) it reaches the plunger chamber (37), through duct (35), where it is mixed with air from duct (36); this mixture is then aspirated through duct (41), so permitting ready starting of the engine. (Diagram A).

As soon as the engine is started, partially close the starter device. **Diagram B**).

In these conditions a further airflow, from duct (42) leans the mixture delivered by the starter, so permitting normal working with a cold engine.

As the engine warms up, however, this mixture is too rich and in excessive supply, so the starting device must be progressively cut out as the temperature of the engine rises.

With the starting device disconnected, the plunger (37) closes duct (41) stopping the call for mixture. (Diagram C).

## INSTRUCTIONS FOR USE OF STARTING DEVICE

In order to get the best results possible from the starting device, the most important instructions for use are summarised below:

## ENGINE STARTING

**Starting from cold** - Fully insert the starting device. Position « A ». On starting, reduce its degree of connection.

**Starting with engine warm** - Partial insertion of the starting device is all that is needed in this case. Position « B ».

**Putting the vehicle to work** - During warming-up of the engine, even with the vehicle in motion, progressively disconnect the device with successive manipulations so as to have always a supplementary distribution of mixture, sufficient but no more than necessary for normal functioning of the engine. Position « B ».

**Normal running of the vehicle** - As soon as the engine has reached a temperature sufficient for normal running, cut out the starting device. Position « C ».

---

Soc. p. Az. **EDOARDO WEBER** - Fabbrica Italiana Carburatori

Stab. } BOLOGNA - Via Timavo 33 - Telef. 41.79.95 (Italy)  
Works } TELEX: 51119 WEBER BO

Ind. Telegrafico }  
Cable Address } WEBER - BOLOGNA

# 250GT ~ ENGINE

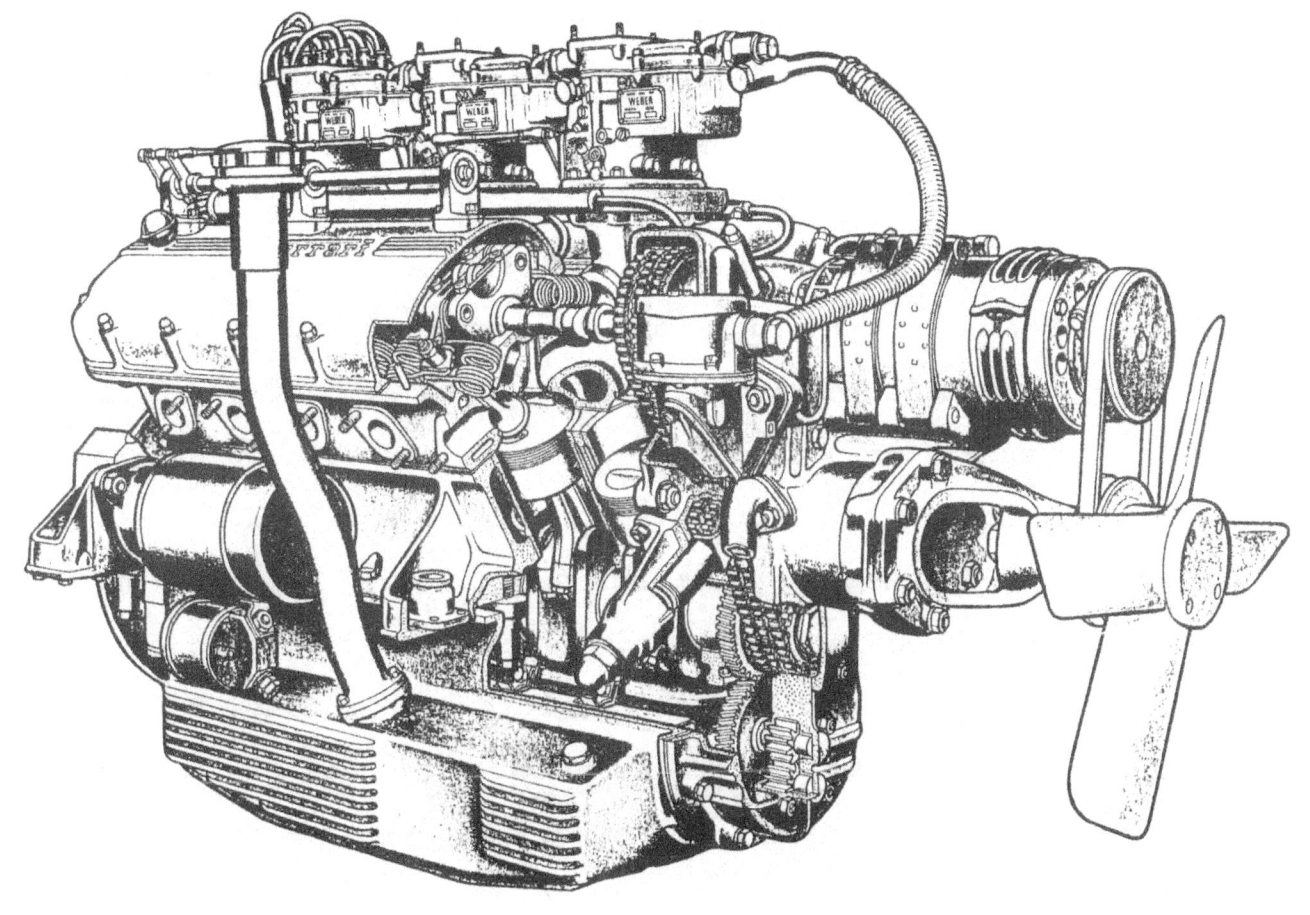

| | |
|---|---|
| CYLINDER HEADS | 56 - 59 |
| CRANKCASE AND CRANKSHAFT | 60 - 62 |
| CAMSHAFTS AND CAM TIMING | 63 - 67 |
| PISTONS AND RINGS | 68 - 69 |
| VALVE ADJUSTMENT | 70 - 72 |
| LUBRICATION SYSTEM | 73 - 74 |
| COOLING SYSTEM | 75 - 78 |

## CYLINDER HEADS

When replacing head gaskets, always use new replacements, and <u>do not</u> apply "Permatex" or other materials to the surfaces. The word "Alto" should be installed facing upward. The block deck should be smooth clean, and dry before installation. Studs and threads should be greased or lubricated before the head is put on. Threads on the stud and nuts must be clean and free from burrs. Washers should be flat and free from burrs; do not substitute washers or nuts other than original replacement parts. Nuts must be brought up to the indicated torques in at least three steps, following the patterns shown. For Example: If the torque specified is 60 lb. ft., on the first pass torque to 25 lb. ft., 45 lb. ft. on the second retorque or pass, and then finally, bring all nuts in the indicated sequences up to their final value. Always consult the owners manual or a <u>qualified</u> expert before attempting to torque the heads. Heads should be torqued cold. Heads should be retorqued or checked after three hundred miles of operation.

Suggested torques are:

| | |
|---|---|
| 250GT, GTE, GTL | 60-62 lb. ft. |
| 275GTB, | 57-59 lb. ft. |
| 330GT, 2+2 | 59-60 lb. ft. |
| 166-212 | 55 lb. ft. |
| 250GT Inside Plug | 55 lb. ft. |
| 275GTB/4 | 59-60 lb. ft. |
| 365GTB/4 | 72 lb. ft. |

## CYLINDER HEAD NUT TORQUE SEQUENCE

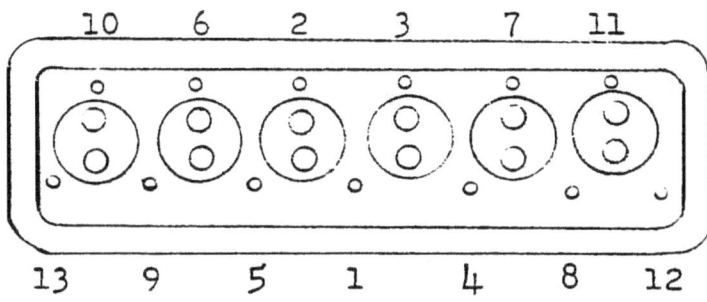

166 - 55 lb. ft.
212 - 55 lb. ft.
250 - 55 lb. ft.

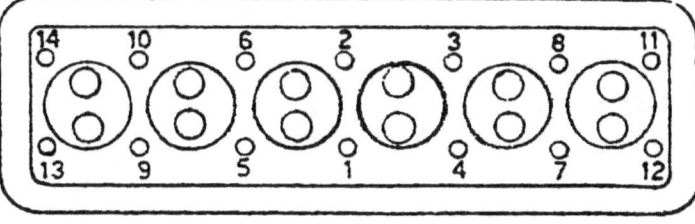

250GTE   60-62 lb. ft.
275GTB   57-59 lb. ft.
330GT    59-60 lb. ft.

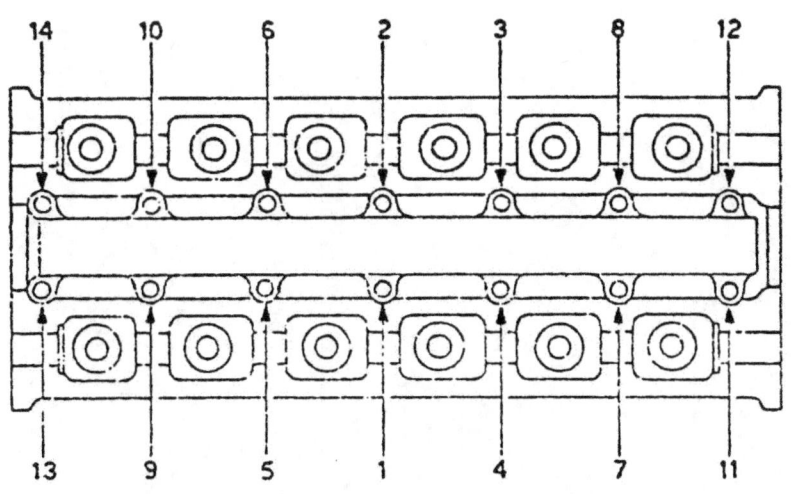

275GTB/4  59-60 lb.
365GTB/4  72 lb. ft.

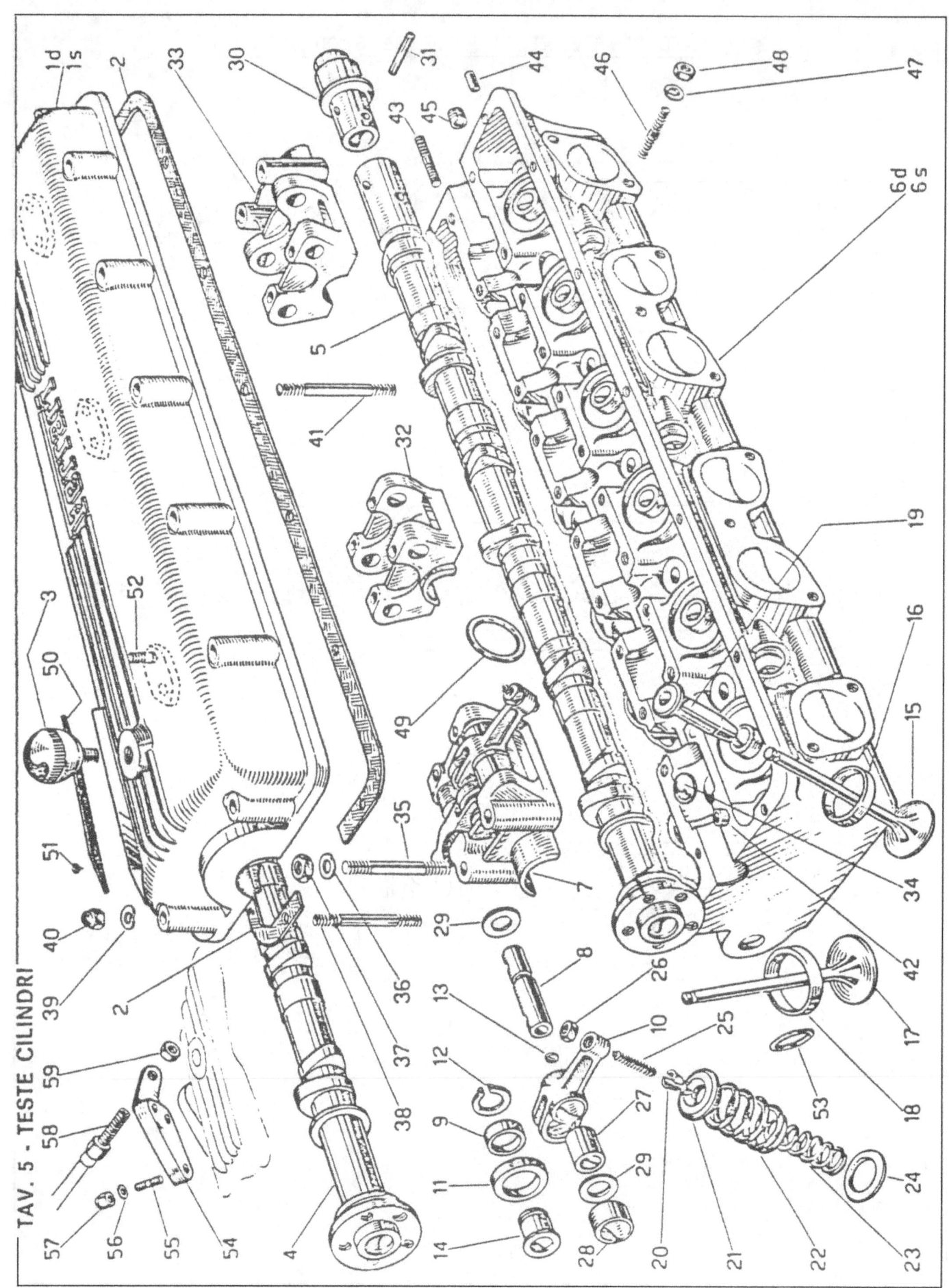

TAV. 5 - TESTE CILINDRI

# MODIFICHE PER INSERIMENTO PARAOLIO SULLE VALVOLE.

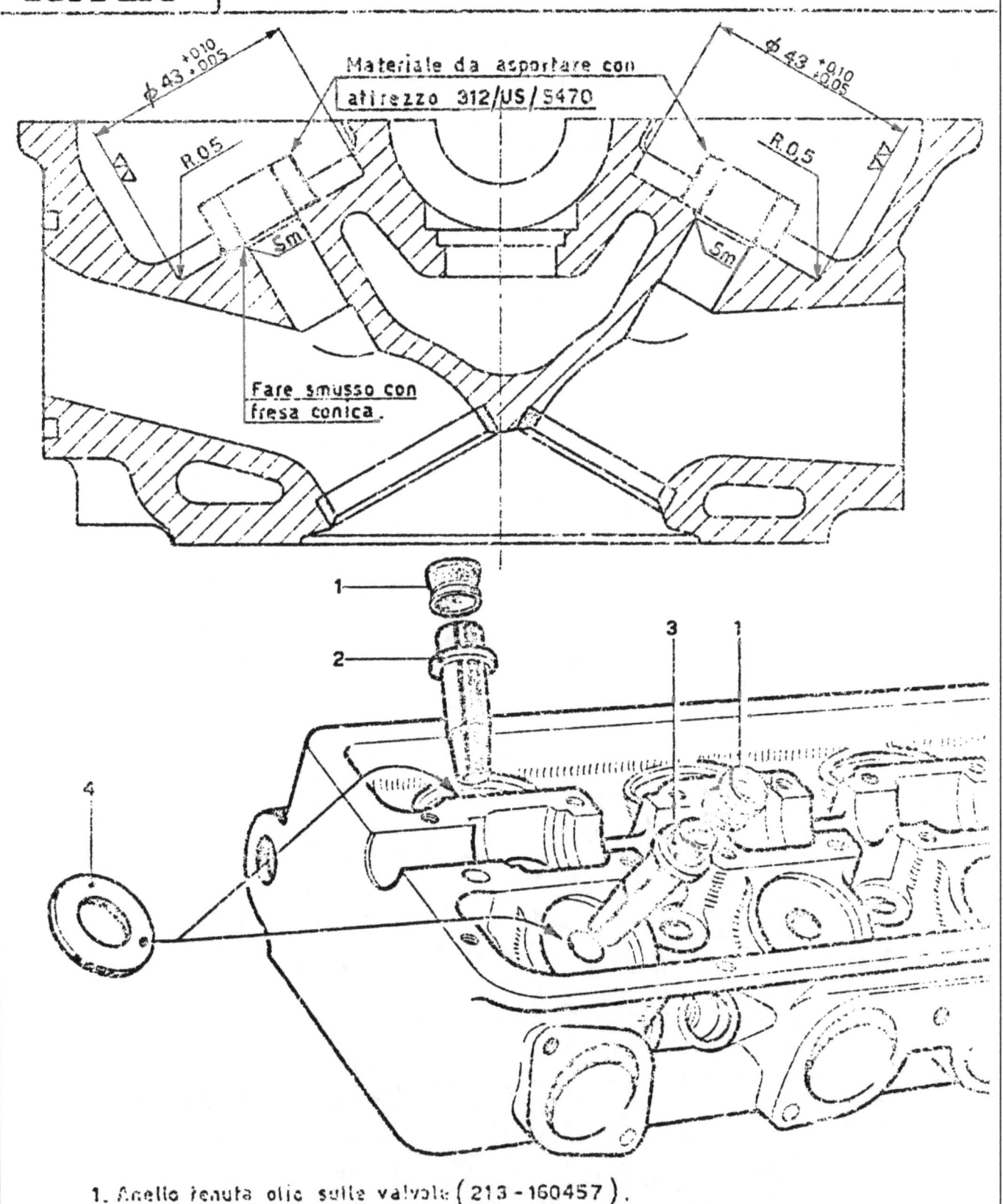

1. Anello tenuta olio sulle valvole (213-160457).
2. Guida valvola aspirazione (213-160413).
3. Guida valvola scarico (213-160414).
4. Piastrina appoggio molle valvola (213-160415).

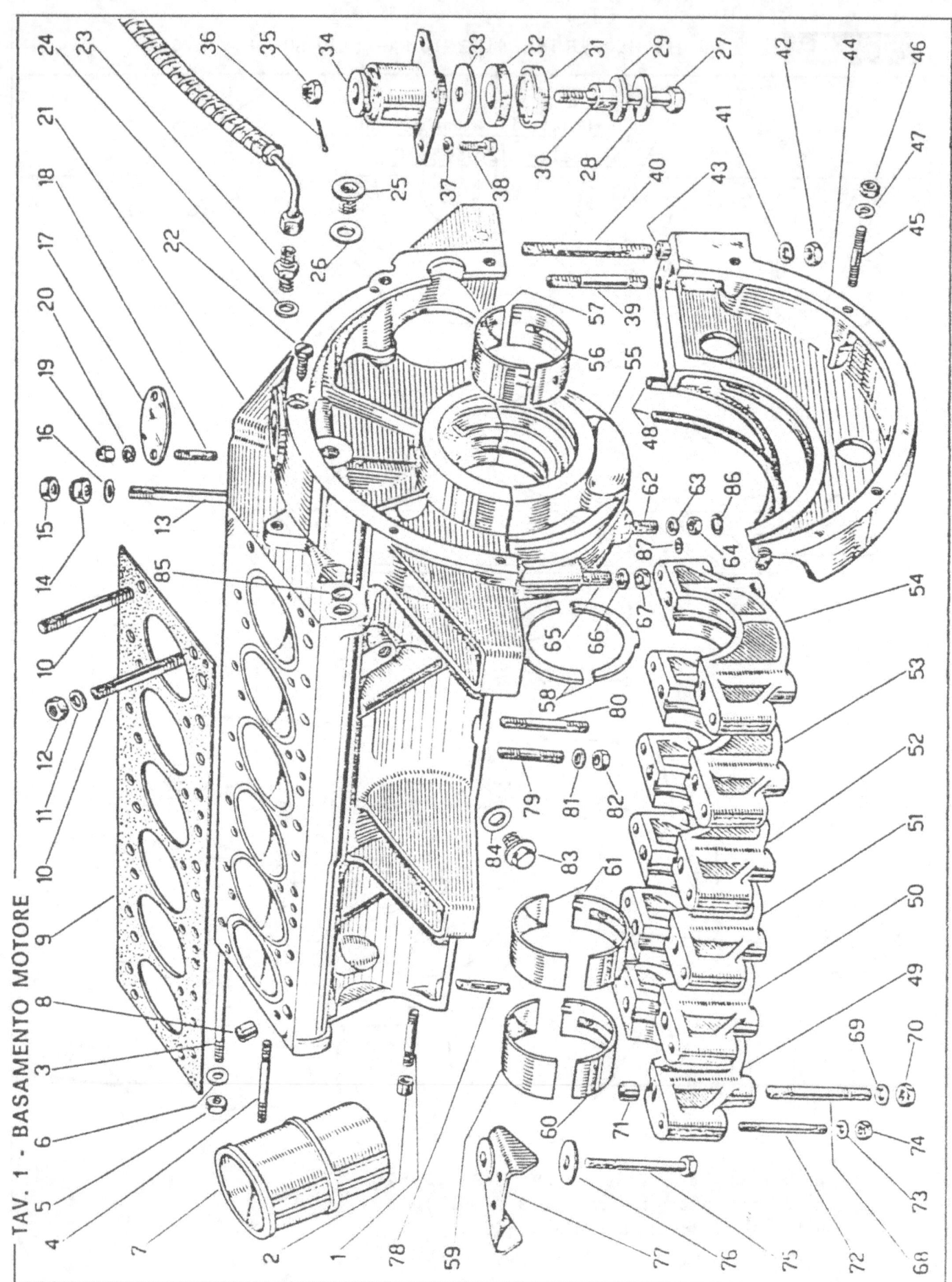

TAV. 1 - BASAMENTO MOTORE

## TABLE OF INSTRUCTIONS FOR REGRINDING CRANKSHAFT JOURNALS
## FERRARI

| MODEL | BEARING JOURNAL | NOMINAL (Inches) | 1st REGRIND .010" | 2nd REGRIND .020" | 3rd REGRIND .030" | 4th REGRIND .040" |
|---|---|---|---|---|---|---|
| 212 Inter 250 mm | Rod | 1.6249" | 1.6149" | 1.6049" | 1.5949" | 1.5849" |
| | Main | 2.1653" | 2.1553" | 2.1453" | 2.1353" | 2.1253" |
| 250GT Inside Plug Up to 1958 | Rod | 1.6249" | 1.6149" | 1.6049" | 1.5949" | 1.5849" |
| | Main | 2.1653" | 2.1553" | 2.1453" | 2.1353" | 2.1253" |
| 250GT 250GTL 250GT/E Outside Plug | Rod | 1.6249" | 1.6149" | 1.6049" | 1.5949" | 1.5849" |
| | Main | 2.3601" 2.3607" | 2.3501" 2.3507" | 2.3401" 2.3407" | 2.3301" 2.3307" | 2.3201" 2.3207" |
| 275GTB | Rod | 1.6235" 1.6241" | 1.6135" 1.6141" | 1.6035" 1.6041" | 1.5935" 1.5941" | 1.5835" 1.5841" |
| | Main | 2.3601" 2.3607" | 2.3501" 2.3507" | 2.3401" 2.3407" | 2.3301" 2.3307" | 2.3201" 2.3207" |
| 330GT | Rod | 1.7172" 1.7177" | 1.7072" 1.7077" | 1.6972" 1.6977" | 1.6872" 1.6877" | 1.6772" 1.6777" |
| | Main | 2.4789" 2.4793" | 2.4689" 2.4694" | 2.4589" 2.4594" | 2.4489" 2.4494" | 2.4389" 2.4394" |
| 340 mm 342 America | Rod | 1.7177" | 1.7074" | 1.6976" | 1.6877" | 1.6777" |
| | Main | 2.3601" 2.3607" | 2.3501" 2.3507" | 2.3401" 2.3407" | 2.3301" 2.3307" | 2.3201" 2.3207" |
| 365GTB/4 Daytona | Rod | 1.7172" 1.7177" | 1.7072" 1.7077" | 1.6972" 1.6977" | 1.6872" 1.6877" | 1.6772" 1.6777" |
| | Main | 2.4789" 2.4794" | 2.4689" 2.4694" | 2.4590" 2.4596" | 2.4489" 2.4494" | - |
| Measurement Log | Rod | " | NOTE: Beyond the above regrinds. the crankshaft must be replaced, or chrome plated and cut to a standard dimension. | | | |
| | Main | " | | | | |

## Thickness of Thrust Semi - Rings

|  | Normal Thickness (Inches) | Oversized Values |  |  |
|---|---|---|---|---|
|  |  | 1st .003" | 2nd .004" | 3rd .005" |
| From | .09291" | .09566" | .09685" | .09763" |
| To | .09094" | .09370" | .09488" | .09566" |

## Crankshaft Mounting Clearances

| PARTS | CLEARANCES IN INCHES |
|---|---|
| Main pin and crank pin permissible ovalization | .0011" |
| Between main pins and bearings (radial clearance) | .0025" - .0033" |
| Between thrust rings and crankshaft (axial clearance) | .003" - .007" |
| Between crank pins and bearings (radial clearance | .0023" - .0031" |
| Between two flanked connecting rods and crank pin shoulders (axial clearance) | .0078" - .0118" |

NOTE:

Check carefully the condition of the Silicone type oil seal before installing it.

Lubricate seal lip with oil and check retainer for eventual damaged spots.

## CAMSHAFTS

When installing camshafts in an overhauled engine, the following steps should be adhered to:

1. Rotate engine until PM 1/6 on the flywheel is exactly on the pointer. Cylinder #1 should be at TDC.

2. Clean all cam bearing surfaces, and heavily lubricate them with SAE20W motor oil. Slacken or remove chain tensioner.

3. Carefully place the camshafts back in their original locations (intake-exhaust L & R) on the head. Mark or tag each cam prior to removal.

4. Place all of the cam bearing-roller assemblies on the cams.*
   NOTE: Each assembly should be returned to the location it was removed from. Check the numbers.

5. Finger tighten the nuts on these assemblies to hold the cam in place.

6. Align** the lightly scribed marks at the front of each camshaft (see diagram) with the arrow stamped in the center of the front cam bearing assembly.

7. Tighten each bearing-roller assembly to the prescribed torque.

8. Place chain tightly over each sprocket.

9. Adjust chain tensioner to take up the slack in the chain.

10. Check to be sure that the arrows still point to the scribed marks on the cam; correct any great error**.

11. Rotate the engines crankshaft two complete turns until PM1/6 is again at the pointer; the cam marks should be at the arrows. Adjust if necessary to correct any great error**.


\* NOTE: Valves must not be actuated (loosen all lash adjustments) when aligning the cam marks for the first time.

\*\* Used or worn chains may not line up exactly - replace chain if great error is noted.

# CAMSHAFT - CHAIN ALIGNMENT — CONTROLLO FASATURA

## TWO CAM

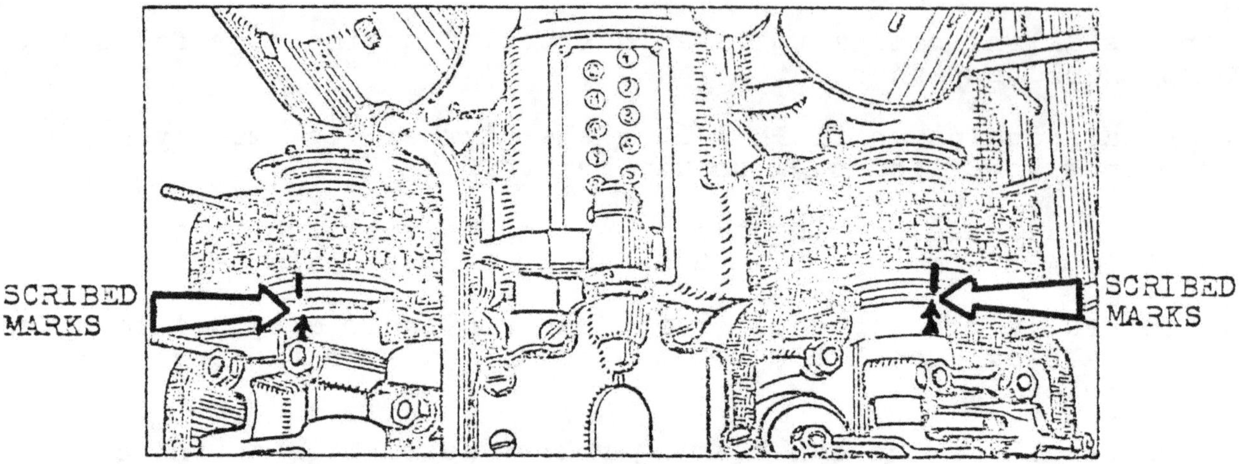

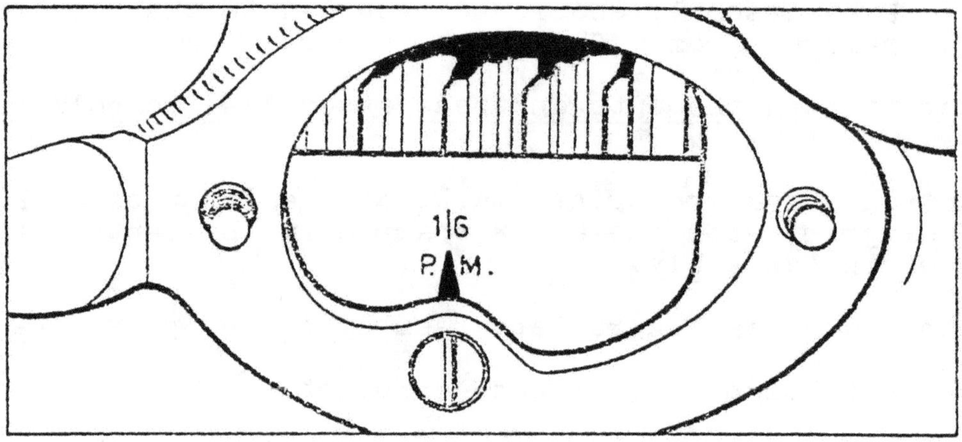

## FOUR CAM
### ALIGNMENT MARKS

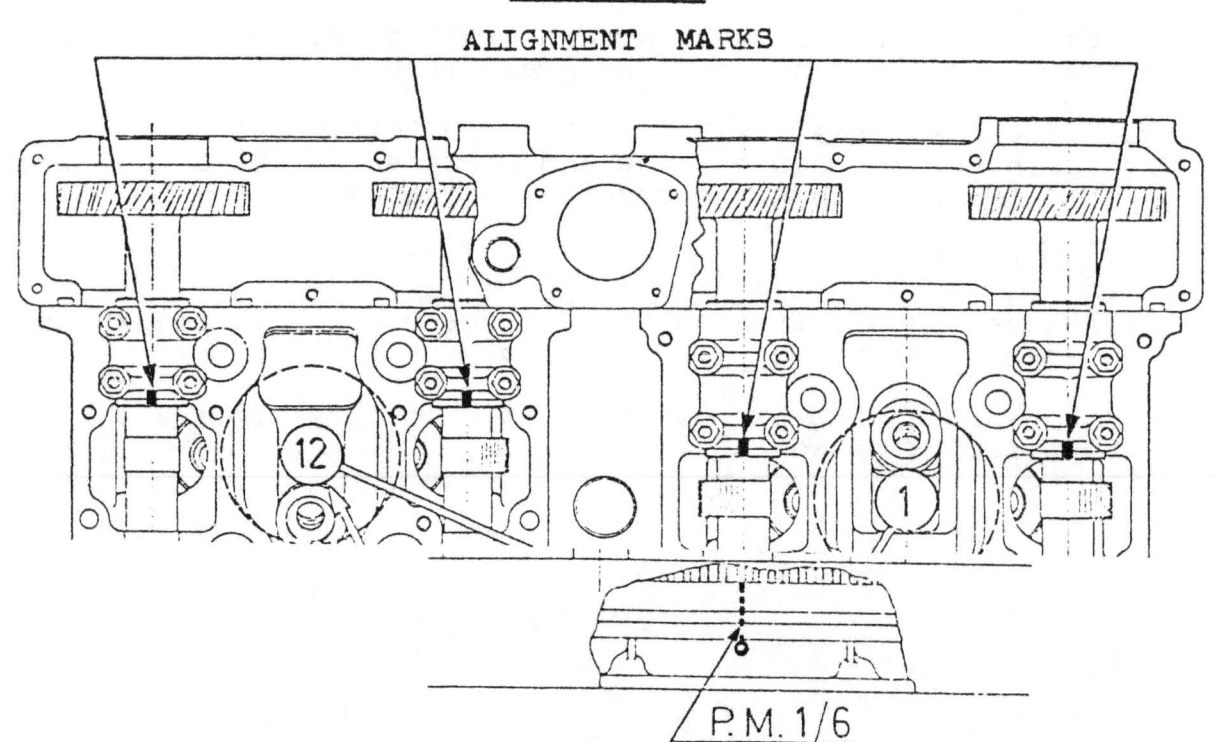

In case of an engine which is being assembled for the first time, or one lacking reference data due to replacement of several parts, proceed as follows:

Turn engine to T.D.C. 1/6 and lubricate the camshaft bearings.

Install both camshafts in such a manner that both cams of the right shaft operating the cylinder No. 6, are turned upwards at the same height, and those of the left shaft operating the cylinder No. 12, also turned upwards, (more specifically, when sitting in the car, cylinder No. 1 is the first on the right hand line closer to the radiator, and cylinder No. 7 is the last in the left hand line closer to the instrument panel).

Install the camshaft drive gears on their proper hubs without fastening the mounting bolts.

Check if the mounting bolts of the left hand gear match with those of the camshaft, in opposite case provide the adjustments as described. Proceed then by fastening the gears with only two mounting bolts and install the rocker arm support caps.

Turn crankshaft 60° rotation-wise up to T.D.C. 7/12, and repeat the same operation for the left hand gear.

Install support caps of rocker arms operating in the same manner as for the right hand shaft.

Adjust valve clearances

Adjust timing chain tension

Install the dial indicator on the mounting flange of clutch housing with the zero mark corresponding to the fixed pointer on the block.

Turn the crankshaft rotation-wise until the rocker arm of cylinder No. 1 is near the opening point of the inlet valve.

Turn gradually the flywheel and hold by hand or with the help of pliers at the exact instant in which the rocker arm roller is no more free to turn.

Read on the dial the position of the T.D.C. 1/6 in relation to the zero mark.

Turn again flywheel, helping with light blows until obtaining the instant in which the exhaust rocker arm roller is free to turn either by hand or with the help of pliers.

Read on dial the number of degrees the T.D.C. 1/6 is retarded in relation to the zero mark.

Example: Suppose the inlet valve starts to open 23° <u>before</u> T.D.C. 1/6, and the closing of the exhaust valve ends 20° <u>after</u> the same T.D.C. 1/6, it will be necessary to advance the inlet opening by 4°, which consequently will retard the exhaust closing also by 4°.

The exact timing should result as follows:

Inlet    Opening, start 27° before T.D.C.
         Closing, end 65° after T.D.C.

Exhaust  Opening, start 74° before T.D.C.
         Closing, end 16° after T.D.C.

- Remove therefore the gear from camshaft and hold chain upwards as high as possible so that it will not slip out of the crankshaft pinion. Rotate gear by 7 teeth in the engine rotation direction and remount gear on proper shaft.

- In case that after this operation the mounting holes will not match properly, it will be necessary to turn the engine in the opposite direction by 4°.

- Fasten gear on shaft as before with two mounting bolts and check readings. If the values obtained correspond with those indicated on table, the timing of the right hand line of cylinders is correct.

- To obtain the left hand line timing, turn engine rotation wise until the inlet rocker arm roller of cylinder 7 is in the start opening position.

- Read the position of T.D.C. 7/12, always in relation to the zero mark on the dial, and to achieve the exact timing, repeat carefully all the operations as described for the right hand line cylinders.

- In case the readings should indicate that it is necessary to retard rather than to advance the inlet opening-start, and consequently to advance the exhaust closing-end, then the positioning of the gear in relation to the timing chain should be effected on the opposite direction to the engine rotation, while that of the engine in the same direction in which it rotates.

- Terminated the timing operation, fasten gears with all mounting bolts and proceed with installing the ignition distributor supports.

The minimum deviation obtainable for the ignition timing, measured on the flywheel if 4°10', therefore the timing tolerance is plus or minus 2°.

The minimum deviation value is obtained from the difference of the angle formed by the 7 teeth of timing gear (148° 10') and that formed by the mounting holes of the same gear about the shaft (144°).

When rotating instead gear and timing shaft by 1 tooth in relation to the chain, the deviation of the flywheel is 21°10'.

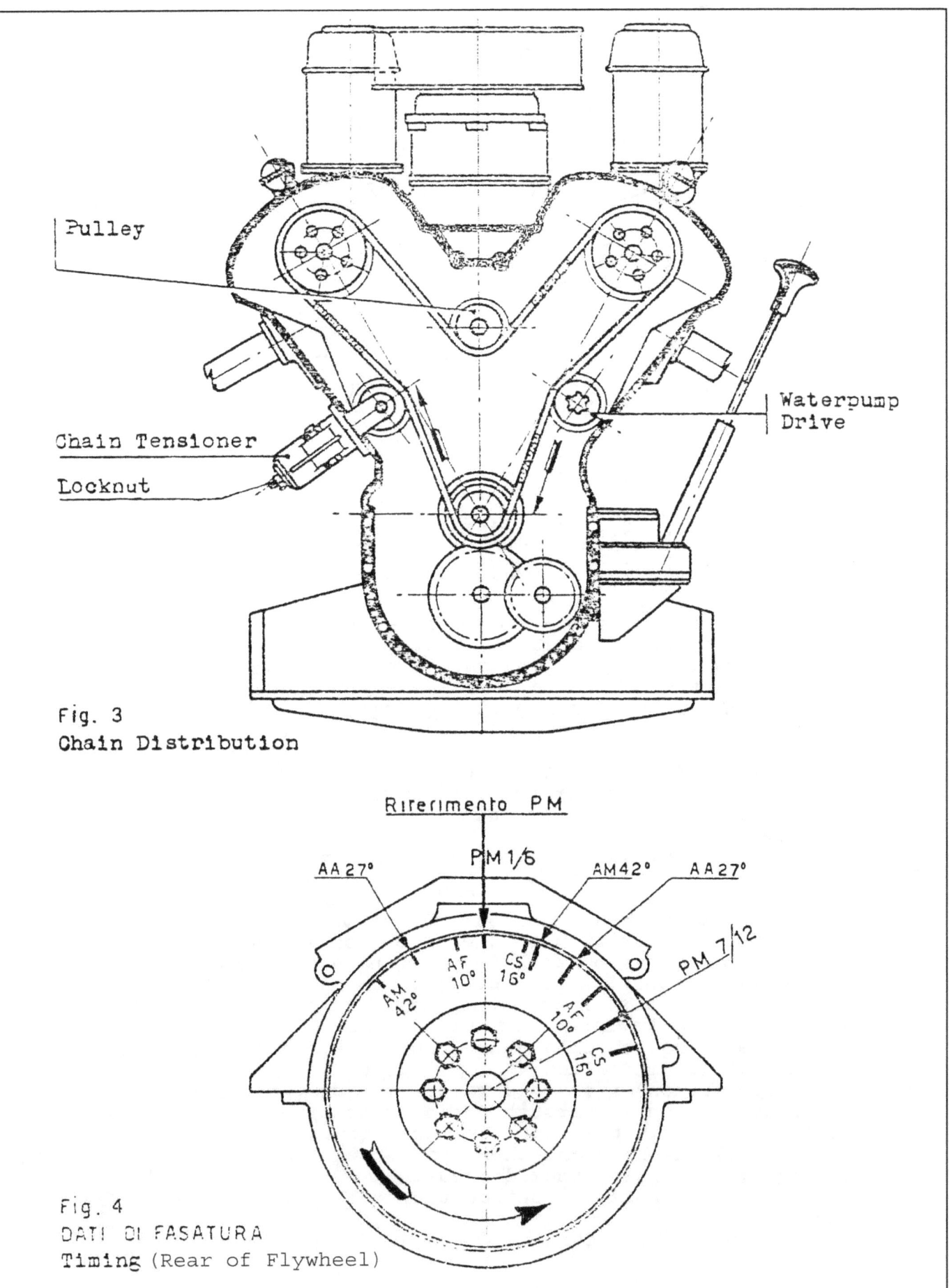

Fig. 3
Chain Distribution

Fig. 4
DATI DI FASATURA
Timing (Rear of Flywheel)

# PISTON RINGS

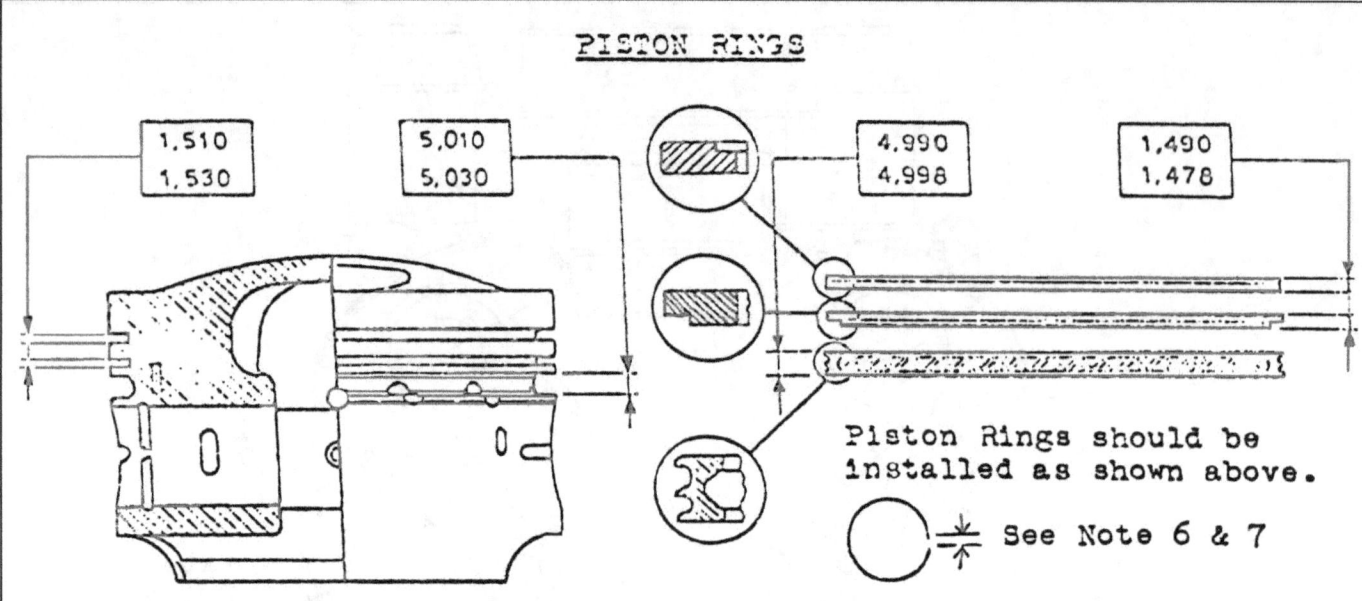

Piston Rings should be installed as shown above.

See Note 6 & 7

### INSTALLATION AS FOLLOWS:

1. Remove the ridge at the top of the bore.
2. Hone the bore rough.
3. Measure the bore at the top, center, and bottom.
4. Machine to a standard diameter if wall damage or heavy taper is measured.
5. Insert a piston ring into the smallest part of the bore and measure the end butt gap.
6. Piston ring end butt gap should be approximately .004" for each 1.00" of bore diameter., i.e. 3" dia. bore, Gap = .012".
7. Piston ring ends (butt gap) should never touch each other in the smallest part of the bore; excessive gap causes loss of pressure
8. Use only new rings for rebuilding engine.
9. The piston must be free from all deposits, especially in the ring groove area.
10. Rings are available in three over sizes from dealers; check the bore diameter and order the proper size.
11. Using a ring expander, install all rings in the order shown in the diagram.
12. Measure the ring to land spacing as shown.
13. Rings must turn free by hand in the grooves.
14. Space the ring end openings 120° from each other.
15. Soak the piston, rings and bore with 20W oil.
16. Each piston must return to the bore from which it was removed.
17. Compress the rings on the piston using a ring compressor.
18. Align the pistons in their original position in the bore.
19. Slowly tap the piston into the bore from above (on late engines) with a wood or rawhide hammer or block.
20. Take care as not to allow the connecting rod to touch the crankshaft (if installed).
21. Check for broken or chipped material from the rings in the bore, after the piston is through the compressor.
22. Replace any rings that chip on insertion.
23. Heavily oil the bore and check for freedom of piston movement in the bore; no binding should occur.
24. Do not push the pistons out of the bore as the rings may be damaged.
25. Repeat for all cylinders.

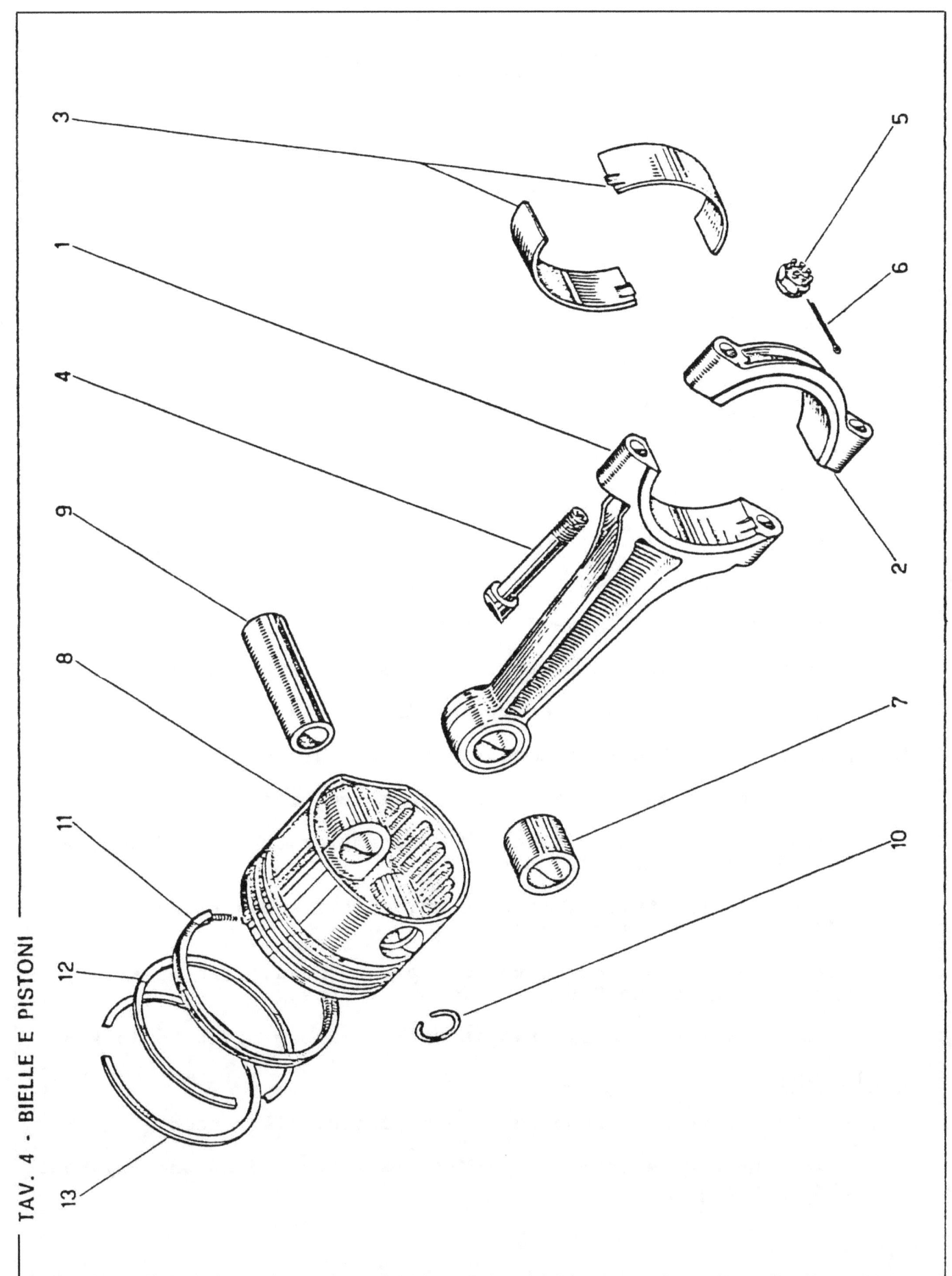

TAV. 4 - BIELLE E PISTONI

## VALVE CLEARANCE ADJUSTMENT
### (250-330 SOHC)

Every 6,000 miles, check the valve lash clearances and if necessary, reset them to the recommended clearance.

### REQUIRED MATERIAL:

    10mm Wrench or Socket

    11mm Box Wrench

    Small Adjustable Wrench

    .006" (.15mm) Feeler Gauge

    .008" (.20mm) Feeler Gauge

    Small Piece of Thin Sheet Metal

    Cam Cover Gaskets (if required)

### CAM COVER REMOVAL:

1. Remove all acorn nuts from the cam covers.
2. Remove all flat washers from the studs.
3. Loosen, (at least half way) all nuts on the chain covers.
4. Remove the bolts that secure the distributor to the cam cover.
5. Lift the ignition wire looms away from the cam covers.
6. Tap the chain covers to be sure they are loose.
7. Remove all carburetor linkages by snapping the clip from the ball joint on the rod.
8. Remove the throttle cable connection at the cam cover. On 330s, remove the air cleaner.
9. Carefully lift the cam cover off of the head, lift upward by the black knobs on the cover (do not force).
10. If the cover hesitates, tap it with a rubber malet to loosen.

### ADJUSTMENT:

NOTE: The engine must be cold when adjusting clearances.

1. Crank the engine over until PM1/6 is indicated at the flywheel timing mark.

## VALVE CLEARANCE ADJUSTMENT (continued)

2. Intake and exhaust valves on cylinder 1 and 6 should be closed, and can be checked or adjusted at this point.

3. Check the clearnace on a cold engine; the gaps for a 250 or 330 are:

    .006" on intake
    .008" on exhaust

4. The intake valves are the ones closest to the carburetors; the exhaust valves are the ones just above the exhaust pipes.

5. To adjust the clearances, loosen the 11mm jam nut on the top of the rocker arm, insert the correct feeler gauge between the valve stem and the adjuster. Turn the adjuster screw with a small adjustable wrench until the gauge can be removed with a slight pull (a small pressure on the feeler guage is o.k.)

6. Tighten (securely) the jam nut - taking caution not to disturb the setting of the adjuster screw. Check the clearance again to determine if the adjuster screw had been moved, repeat if it is not correct. (Holding the adjuster screw while tightening the jam nut will prevent this problem). Adjust one valve at a time, but complete both adjustments on each cylinder before turning engine over.

7. With a flashlite, observe the position of the cam lobe on the next cylinder to be checked. Rotate the engine until the lobe is pointing away from the rocker arm; the valve is fully closed at this time. Each valve may be adjusted on each cylinder in this manner. (A pushbutton switch connected from the starter fuse box to battery will speed up cranking the engine over each time. Intake and exhaust valves have different clearances; do not mix the measurement up.

7a. Another method of determining if the valves are closed is to crank the engine through the normal firing order and adjust each valve lifter when the piston is at TDC. Starting with cylinder #1 with PM 1/6 on the flywheel mark, crank the engine slightly to the next cylinder in the firing order, then check the clearance. The firing order on 250 and 330 cars is as follows: 1-7-5-11-3-9-6-12-2-8-4-10. (When a cylinder is at TDC in the firing cycle, both valves of that cylinder are closed and adjustments of the clearances on the intake and exhaust valves can be made at this time.)

8. With either method 7 or 7a, recheck each valve again to double check the clearances. Also, double check the tightness of the jam nuts.

9. The engine can be run with the cam covers off to check for any noisy adjustments. Do not run the engine at high RPM or oil may splash about the engine compartment.

## VALVE CLEARANCE ADJUSTMENT (continued)

10. If readjustment of a particular valve clearance is indicated, wait until the engine is cold before setting.

REASSEMBLY:

1. If the large cover gasket appears dry, cracked, decomposed, or broken in any way, replace it with a new one. Remove the chain covers and lift the old gasket off (avoid damaging the front oil gasket). Do not drop any pieces into the engine. Clean the area under the gasket and remove all adhesives if they were applied. Install a new gasket, cutting off (carefully) the small section of gasket over the distributor drive. The oil gasket between head and chain cover should be carefully inserted through the small punched squares in the gasket. Push the gasket down on the head, do not use adhesives to hold the gasket.

2. Clean the gasket area on the cam case (and chain cover if a new gasket is to be installed) removing all adhesives or traces of gasket material. Check the oil drain holes for blockages; clean them if it is required.

3. Install the chain cover if it was removed. Place a thin sheet metal piece against the oil ring, holding it against the chain cover.

4. Using the sheetmetal piece to hold the oil gasket against the chain cover, carefully slide the cam cover down on the head. Slowly remove the sheetmetal, apply oil from above if it is difficult to remove. (Do not use adhesives to hold gaskets in place.)

5. Place the flat washers on the studs, then place the four spacers on the longer studs. Fit the ignition wire harness over the proper studs.

6. Install the 6mm acorn nuts on all of the studs and run them down finger tight. Be sure the covers are down flat on the heads.

7. Torque each nut down to about 6 lb. ft. Tighten nuts in order from the center to the ends of each cover.

8. Install the distributor securring bolts and tighten down.

9. Install the carburetor linkages and throttle cable. Do not adjust the throttle cable taut, or it may change the idle setting. Install the air cleaner if it was removed.

10. Start the engine and check for oil leaks and loose cover nuts.

# OIL SYSTEM

## PARTS DESCRIPTION

1. Oil Pump
2. Pressure Adj. Cover
3. Fram PH 2815 Filter
6. Fram PB-50
20. Pick Up Screen
60. Lock Nut

## ENGINE LUBRICATION

A gear pump is used to provide oil to the engine. The <u>maximum</u> oil pressure can be adjusted by means of a by-pass valve located on the oil pump. To increase the oil pressure, remove the large brass cover nut (#2) with a 15/16 or 24mm socket, slacken the thin lock nut (60) just below the cover nut. To increase the maximum oil pressure, screw the large slotted screw down (clockwise), to decrease turn counter clockwise.

## ENGINE LUBRICATION (continued)

This adjustment will not increase the oil pressure at low speeds, but rather limit the maximum pressure to a safe level. Pressures above the recommended limits can cause damage to the engine. Should the pressure fall below 38 to 45 PSI at high RPMs, do not continue running the engine for long periods of time. Recommended pressures are:

| ENGINE SPEED | OIL TEMPERATURE | OIL PRESSURE(PSI) | OIL PRESSURE(METRIC) |
|---|---|---|---|
| 6600 RPM | 212°F/100°C | 80PSI (Normal) | 60 (Normal) |
| 6600 RPM | 212°F/100°C | 55PSI (Minimum) | 40 (Minimum) |
| 6600 RPM | ALL | 38-45PSI (Danger) | 35-40 (Danger) |
| 800 RPM | (Depends on oil) | 14-21PSI (Normal) | 10-15 (Normal) |

NOTE: Never open the throttle full out until the oil reaches a temperature of 140°F (60°C).

Clean filter screen (20) if pressure is low. The screen should be cleaned at least every 30,000 miles. If full oil pressure is not restored, check for a damaged oil pump, clogged filter or lines, excessive bearing clearances, broken oil lines, or a defective pressure guage.

## MAINTENANCE

Change oil and filters every 3,000 miles (sooner in dusty conditions). Lubricate the rubber gasket on the filter before installing, and do not over tighten.

## COOLING SYSTEM

### MAINTENANCE:

Every 300 Miles — Check the level of water in the radiator, top up 3/4" from cap seat with distilled water if required.

Every 3,000 Miles — Check electric fan drive for operation, check or set gap (between electric fan clutch faces) to .014".
Fan(s) should activate at 185°-194°F.
Check dynamo belt tension, set tension to 3/8" deflection with 9 lb. pressure on belt.

Every 12,000 Miles — Check water pump glands and ball bearings, replace if leaky or noisy. Test radiator cap to 5 PSI relief pressure. Check all hoses and heater circuits for leaks. Drain and flush entire cooling system. Check fan armature brush and slip ring for wear. Replace fan belt.

### SPECIFICATIONS:

| | |
|---|---|
| Radiator Cap | Type R9 - 5-6 PSI Relief |
| Thermostat | 185°F Opening |
| Peugeot Switch | 185°F Opening, 167°F Closing |
| Fan Belt | 3/8" HD Automotive Type |
| Cooling System Capacity | 12 Quarts* |
| Drain Plugs | Two in block, one in radiator |

### ANTIFREEZE:

| Temperature | Quantity of Antifreeze* | Percent |
|---|---|---|
| Down to + 14°F | 5 Pints | 22% |
| Down to + 5°F | 6 Pints | 28% |
| Down to - 4°F | 8 Pints | 34% |
| Down to - 22°F | 10 Pints | 44% |

*Quantities indicated are for permanent type antifreeze in models 250GT and 330GT without airconditioning.

## Cooling system

The engine cooling is assured by the circulating water provided by a centrifugal pump.

The following parts make up the cooling system:

- Centrifugal pump.

- Radiator block with three sets of vertical tubes and an air-tight cap with a relief valve.

- A electric fan, three blade Peugeot type, with an automatic switch control.

- The thermostat located between the engine water outlet and radiator.

- A thermo-contact which operates as a switch control.

## Water pump

The water pump group is attached to the distribution housing. The pump shaft, on one end is slotted and it mounts the pinion which is operated by the camshaft chain, the other end of the shaft is square and it mounts the propeller in bronze, fastened by a nut and washer. The propeller is provided by a special steel washer against which sets the graphitic ring of the seal assemly. The shaft rotates on two ball bearings, of different size, the larger bearing is mounted on the distribution housing, the other on that of the pump body. Between this second bearing and the seal assembly is installed the oil seal ring. A drain hole is provided for eventual water and oil leaks.

## Disassembling the pump

The disassembling operation of the pump group from the car requires some precautions: by removing the complete shaft and propeller group, the control pinion will remain without a proper support and may fall in the bottom of the housing causing at the same time the alteration of the engine timing.

The shaft, nevertheless, is not subject to wear and does not require frequent disassembling.

To dismount the pump body therefore it is sufficient to remove the nut and washer which mount the propeller to the shaft, then remove all nuts which fasten the pump body to the distribution housing and replace the hot water inlet tube of pump with a flanged threaded fitting pointing to the threaded end of the shaft. By introducing it into the shaft, the pump body will come out without any danger of extracting the shaft.

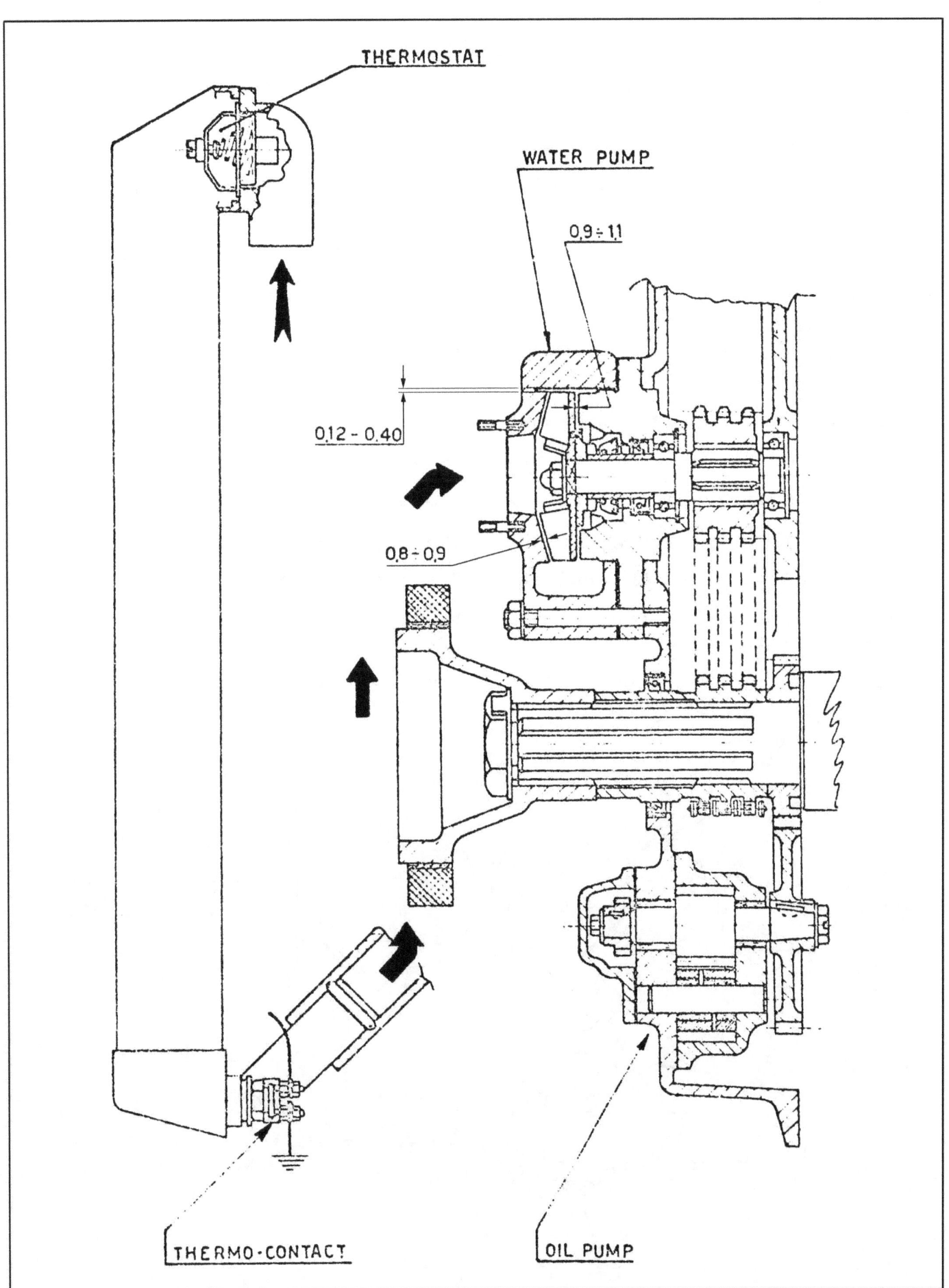

## PRECAUTIONS

To prevent freezing, drain the water from the cooling system if antifreeze is not used.

Flush system before adding antifreeze.

Keep antifreeze in winter and summer to prevent oxidation of the cooling system.

Do not allow water temperature to rise above 203-212°F (95-100°C).

Do not use pressure cap with relief pressure higher than the specified (6 PSI).

If Peugeot electric fan will not energize, tighten the three adjusting screws until the fan functions.

Carry a spare fan belt for emergencies.

Do not operate vehicle without coolants.

Do not operate vehicle for long periods without a fan belt.

Do not release pressure cap when water temperature is above 190°F.

Do not add water to the radiator if engine is hot.

Do not over-tighten fan belt as this may cause damage to the dynamo bearings.

Always use soft or distilled water in the cooling system.

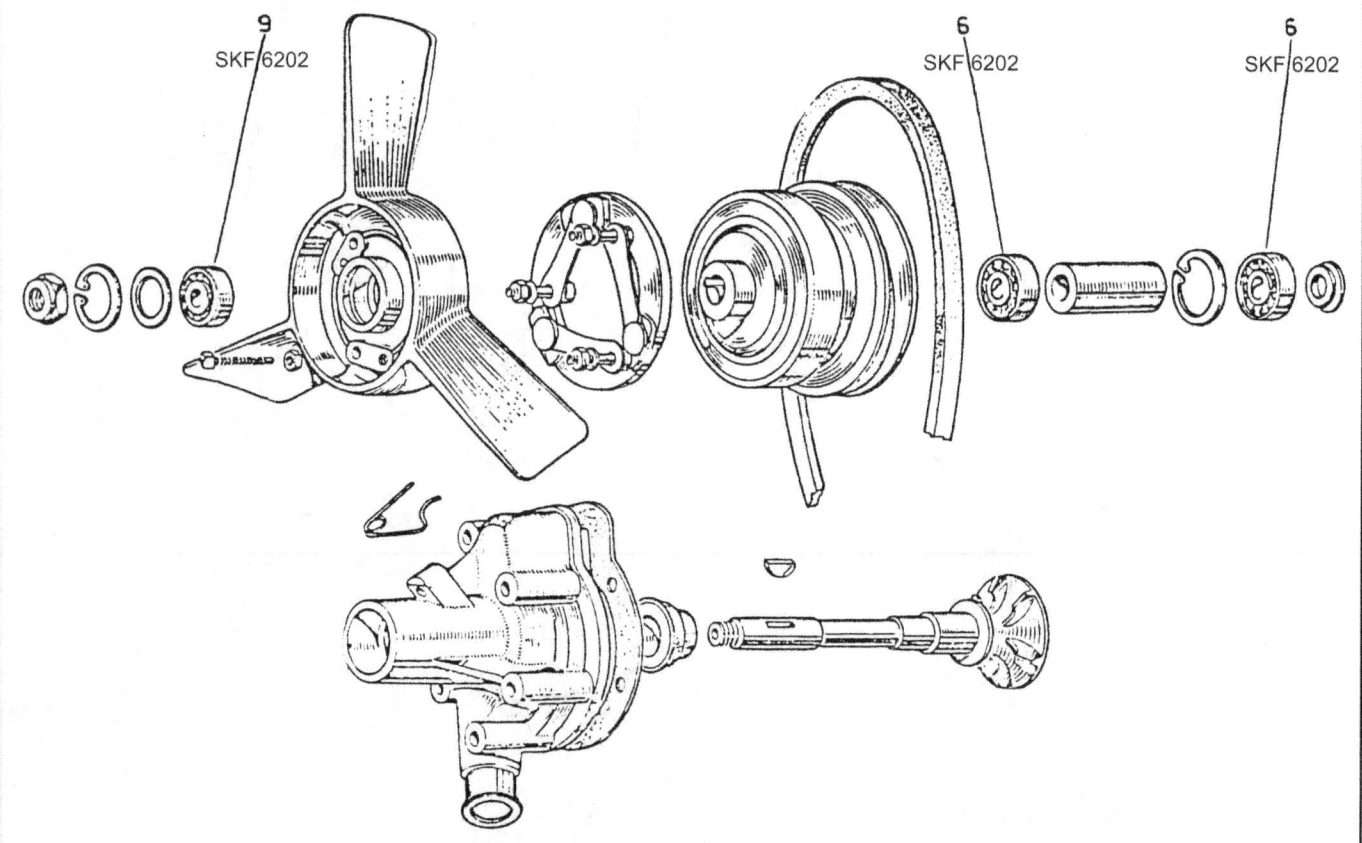

# 250GT ~ GEARBOX AND CLUTCH

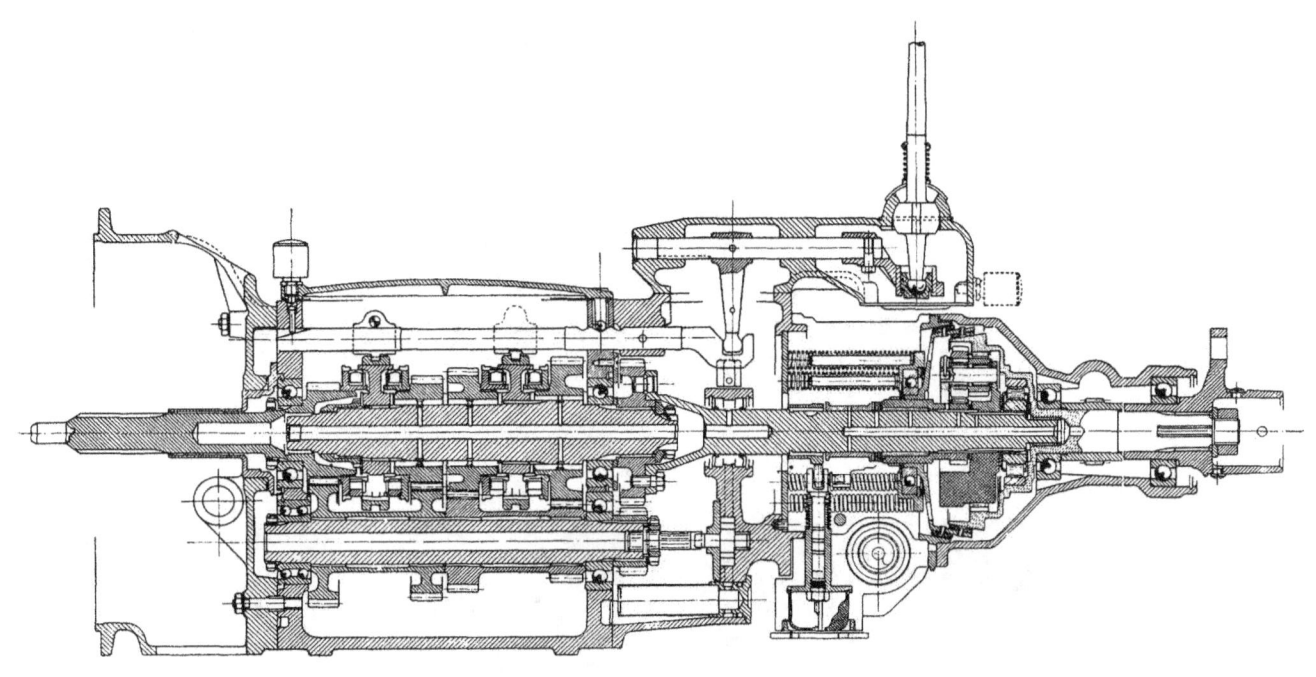

Longitudinal section of gearbox and overdrive.

| | |
|---|---|
| GENERAL DATA | 80 |
| GEARBOX AND CLUTCH REMOVAL AND REPLACEMENT | 81 - 84 |
| CLUTCH SPECIFICATIONS AND ADJUSTMENTS | 85 - 87 |
| SHIFT LEVER BUSHING REPAIR | 88 |
| OVERDRIVE | 89 - 108 |

| Gear ratio | 1st. speed | - 1 : 2.536 |
| | 2nd. speed | - 1 : 1.777 |
| | 3rd. speed | - 1 : 1.256 |
| | 4th. speed | - 1 : 1 |
| | 5th. speed (overdrive) | - 1 : 0.778 |
| | Reverse | - 1 : 3.218 |

adjustment to the gearbox must be performed by a workshop.

check that the oil level is still 1 cm. below the filling hole.

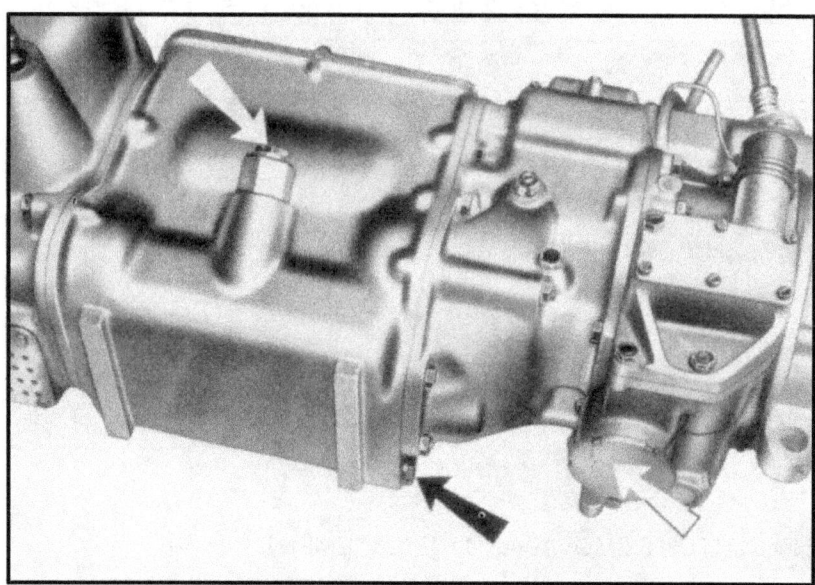

Fig. 42 - Filler and drain plugs.

# GEARBOX & CLUTCH REMOVAL

Models: 250GT, 330GT, with engine mounted gearboxes.

## GEARBOX REMOVAL

1. Remove both (front) seats.
2. Remove carpets and gearshift knob.
3. Unbolt transmission tunnel from fire wall and floor.
4. Lift the tunnel out, carefully clearing the shift lever.
5. Unbolt the rear drive shaft flange at the universal joint.
   Note: On some models, it may be easier to disconnect the front universal joint.
6. Remove the drive line from the transmission, about 3" separation is sufficient to clear the rearward motion of the box.
7. Unbolt the transmission mounting(s) located directly under the tail stock. Some gearboxes, and those with overdrive will have additional mounts to remove.
8. Drain the transmission fluid on gearboxes fitted with overdrive; this is optional on straight gearboxes and is intended to reduce the total weight.
9. Remove all bellhousing bolts.
10. Remove the clutch linkage and speedometer cable.
11. Disconnect any electrical connections, such as backup lamp and overdrive control wires.
12. Place the shift lever in neutral.
13. Place a small jack under the rear of the engine.

## GEARBOX REMOVAL (continued)

14. Note: The gearbox is extracted through the inside of the vehicle.

15. Rock the box from side to side to be sure that all attachments have been removed.

16. Place wooden blocks under the front of the gearbox so it will not drop to the ground when disengaged from the engine.

17. With an assistant, slide the gearbox rearward away from the engine, a slight rocking motion may help free it up.

18. Carefully slide the spline pilot shaft all of the way out of the engine; the gearbox should be completely free of the vehicle.

19. Lift out of the vehicle and place on wooden supports, do not drop the gearbox on hard surfaces. This could crack the housings. To prevent damage to the splines, cover them with rags.

## CLUTCH REMOVAL:

1. With the gearbox removed, the clutch assembly is completely accessible at the rear of the engine.//
2. Using a socket wrench, unbolt the six pressure plate securing bolts. Loosen each bolt a few turns at a time, so that the pressure plate pushes away from the flywheel evenly without binding. (A six point socket is recommended. They will prevent stripping of the bolt heads)
3. The pressure plate and disc are now free to be removed. Slip the throw out bearing off of the shaft.
4. Inspect the flywheel and pressure plate for damage or warpage. Have a qualified expert rebuild or replace all damage materials.
5. Inspect the pilot bearing in the rear of the crankshaft. Replace or re-grease as required.

## CLUTCH INSTALLATION:

1. If the flywheel has been removed, remount it to the rear of the crank shaft, aligning the locating pin carefully.
2. Install all of the original flywheel bolts (do not substitute lower grade bolts) and keeper straps. Torque flywheel bolts to 35lb. ft. Recheck all bolts, and pein keeper straps over.
3. Note: If exessive material was cut from the pressure plate, the flywheel bolt heads may have to be ground down. Check with the dealer for additional information.
4. Using a clutch alignment shaft (most auto supply stores have adjustable sets) place the disc and pressure plate against the flywheel.

## CLUTCH INSTALLATION (continued)

5. Engage the alignment tool into the crankshaft pilot bearing and center the disc under the pressure plate.
6. Install all pressure plate screws, and slowly tighten each a few turns at a time until tight. Torque to 30 lb. ft. Recheck all torques again, and remove alignment tool.

## GEARBOX INSTALLATION

1. With the gear lever in neutral, set the box on the wooden support ready to insert onto the engine.
2. Lift the gearbox into the rear of the engine. A small amount of side motion must be applied to seat the spline in the clutch.
3. Slide the gearbox all the way up to the engine so that the studs protrude through the bellhousing.
4. Install the bellhousing nuts and lockwashers, and torque them to 20 lb. ft.
5. Connect all electrical and speedometer cables to the gearbox.
6. Connect the clutch linkage and adjust to allow correct freeplay. (1-5/8")
7. Bolt the rear transmission mounts down and apply safety wire where holes are provided.
8. Remove jack from under engine.
9. Connect the drive shaft, torque and safety wire all bolts.
10. Fill the gearbox, and overdrive if fitted, with fresh lube.
11. Replace the transmission tunnel, gearshift knob, carpets & seats.
12. Check the clutch and gearbox for proper operation, re-adjust the clutch linkage if necessary.

# CLUTCH

## DESCRIPTION

250GT    - 8-3/4" Diameter Fitchel & Sachs, part no. 1861-084-002

330GT2+2 - 9.5" Diameter Textar, part number 505/17

330GTC   - 9.5" Diameter Borg & Beck

365GTB/4 - 10-1/2" Diameter Borg & Beck, part number BB9/445A

## MAINTENANCE

Every 500 miles - Check free play.

Every 1,000 miles - Adjust linkage and pedal free play.

Every 2,000 miles - Grease linkage.

Every 5,000 miles - Bleed brake fluid if hydraulic actuated type.

Every 10,000 miles - Check disc and pressure plate for damage or wear.

Every 15,000 miles - Rebuild hydraulic cylinder.

20,000 miles - Replace clutch disc.

20,000 miles - Recondition flywheel and pressure plate.

20,000 miles - Replace throw out bearing.

## ADJUSTMENT

Pedal free play - 1/3/8" to 1-9/16"

Pressure plate finger clearance - 3mm

Actuating arm angle - 10°

## PRECAUTIONS

Do not disengage the clutch rapidly when vehicle is at rest.

Do not rest foot on clutch pedal while driving.

Do not keep clutch disengaged for long periods of time.

Do not tow trailers or push other vehicles.

## CLUTCH ASSEMBLY

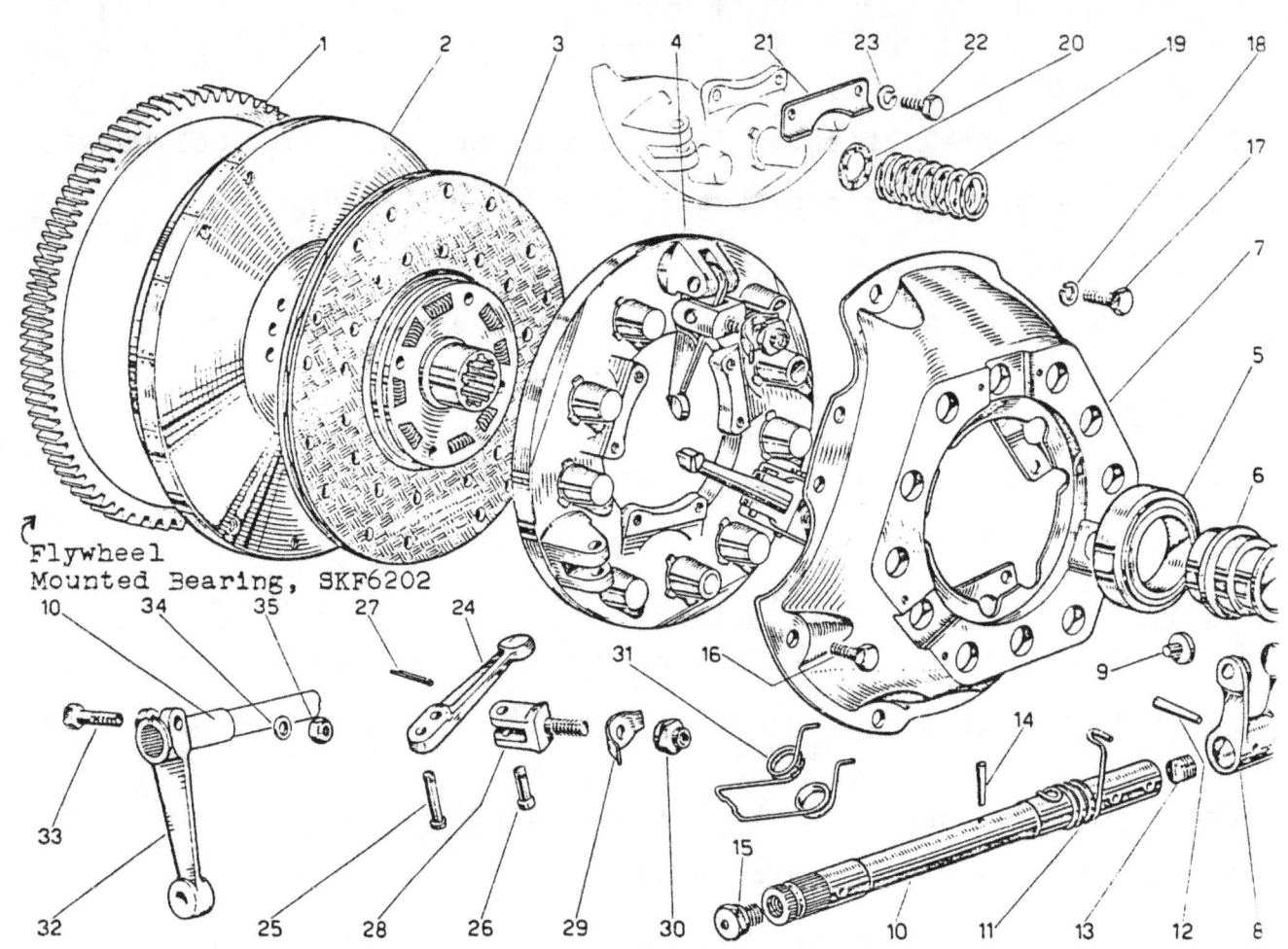

Flywheel Mounted Bearing, SKF6202

DESCRIPTION:
1 - Starter Ring
2 - Flywheel
3 - Clutch Disc
4 - Pressure Plate Face
5 - Throw Out Bearing
6 - Sleeve
7 - Pressure Plate Cover
Pilot Bearing SKF6202

TORQUE SETTINGS: Flywheel Bolts Torque to 35 lb. ft.
#16 Pressure plate bolts torque 35 lb. ft.

ADJUSTMENT: Pedal Free Play 1-3/8" to 1-9/16"

| # | PART | MODEL USED ON | MANUFACTURE | MANUFACTURER'S PART NO. |
|---|---|---|---|---|
| 3 | Clutch Disc. | 250GT | Fitchel & Sachs | 1861-084-002 |
| 4 | Pressure Plate | 250GT | Fitchel & Sachs | -- |
| 3 | Clutch Disc. | 330GT2+2 | Textar Borg & Beck | 505/17 |
| 4 | Pressure Plate | 320GT2+2 | Fitchel & Sachs | -- |
| 3 | Clutch Disc. | 330GTS | Borg & Beck | 9.5" |
| 3 | Clutch Disc. | 365GTB/4 | Borg & Beck | BB9/445A |

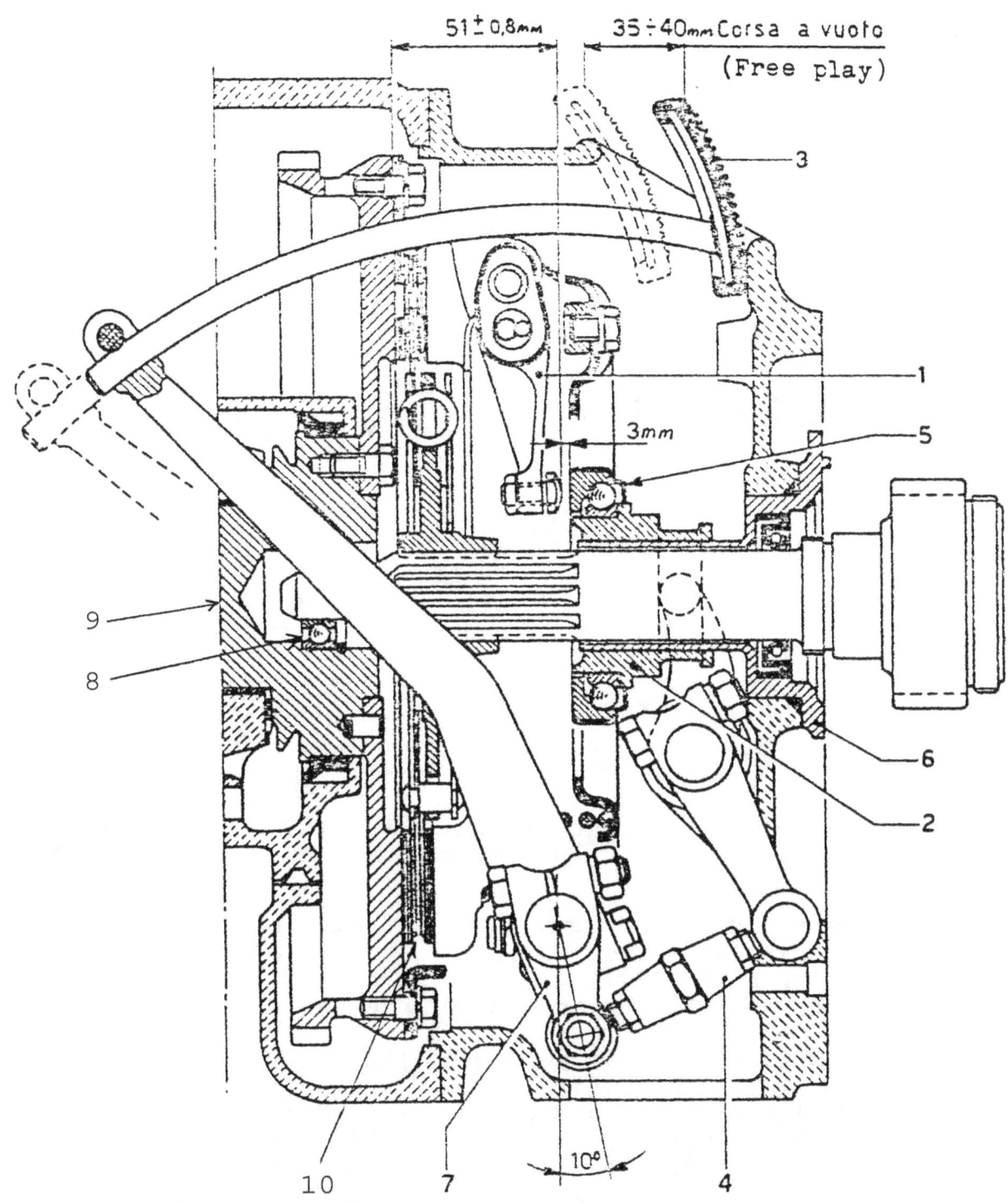

FRIZIONE (CLUTCH)

1. Activating Fingers
2. Sleeve
3. Clutch Pedal
4. Pedal Free Play Adjustment
5. Throw Out Bearing
6. Bellhousing
7. Link
8. Pilot Bearing (SKF6202)
9. Crankshaft
10. Clutch Disc

# REPAIR OF SHIFT LEVER BUSHING ON 250GT-330GT MODELS

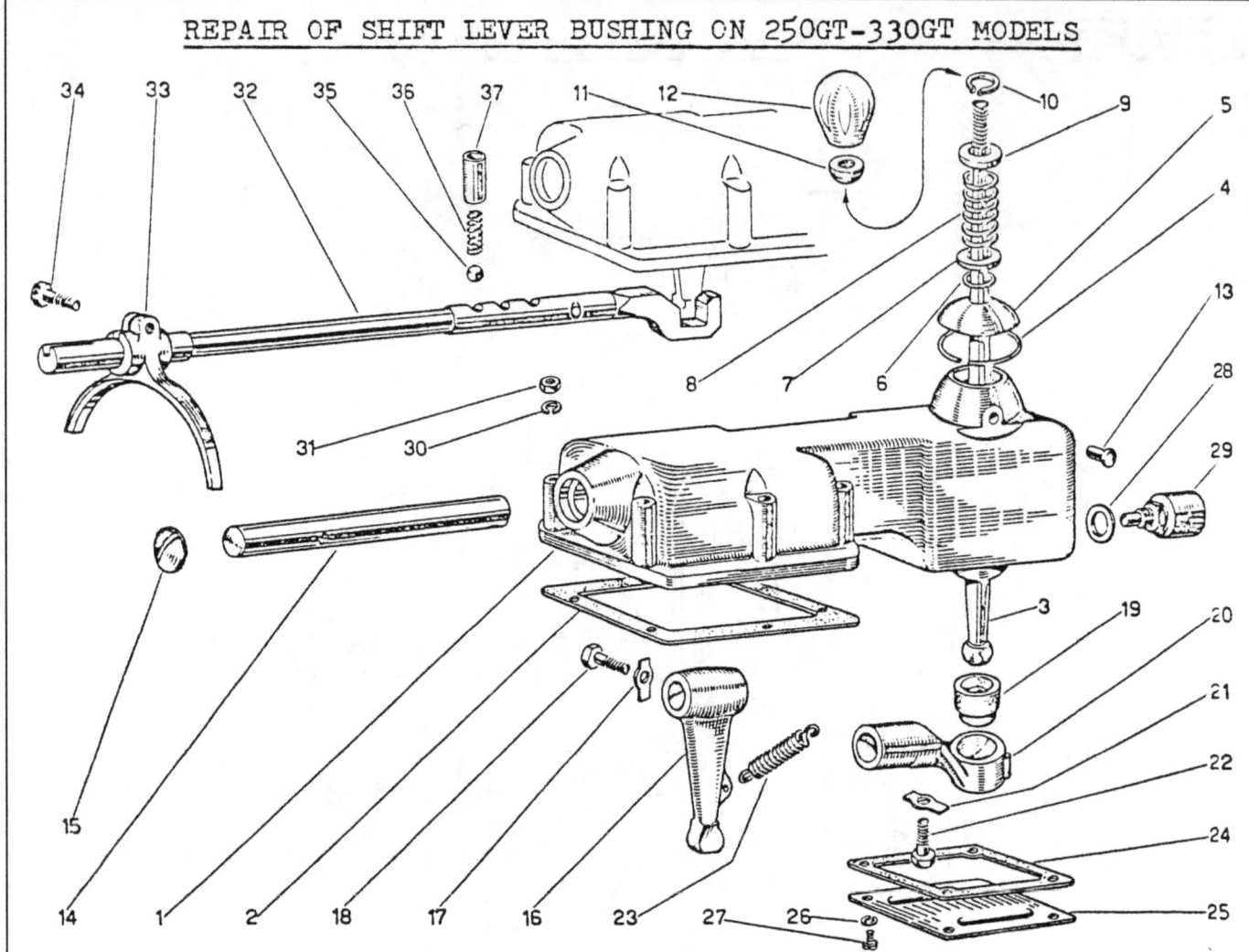

If gear shift lever becomes sloppy and loose, the shift lever bushing (19) may be cracked. This is a common problem, especially under fast shifting.

REMOVAL:
1. Remove gear shift knob (12) and jam nut (11).
2. Remove transmission tunnel cover.
3. Remove housing nuts (31) and lock washer (30).
4. Lift lever housing (1) up and out of gearbox (lever must be in neutral).
5. Remove screws (27) and washers (26), lift cover plate (25) off.
6. Remove bolt (22) and tab (21).
7. Slide shaft (14) forward and pull socket (20) off of shaft.
8. Pull broken or cracked nylon bushing (19) from lever (3).
9. Check spring (23); remove if broken and replace.

INSTALLATION:
1. Grease ball of lever (3) with bearing grease.
2. Force new bushing (19) Ferrari pt. #55467 on ball of lever (3).
3. Push shaft (14) forward.
4. Push socket (20) over bushing (19).
5. Push shaft (14) back into socket (20) until bolt hole lines up.
6. Install bolt (22) and tab (21) bending tab up after tightning.
7. Grease all moving components well and install cover (25).
8. With shift lever (3) in neutral polition, install in gearbox.
9. Torque case nuts (31) to approximately 5 lb. ft.
   (Replace gaskets (2) and (24) if they are broken.

# LAYCOCK OVERDRIVE SERVICE MANUAL

## Ferrari
## 250 GTE
## 330 GT MK1

**IMPORTANT**

While this official 'Laycock Overdrive Service Manual' is identified as being specific for the Ferrari 250 & 330 series equiped with a Laycock overdrive, it originaly included repair data for both the 'Type A' and 'Type D' units.

However, as the Ferrari 250 and 350 series only utilized the 'Type A' overdrive unit, the detailed overhaul section for the 'Type D' has been omitted. Consequently, the sequence of illustration numbers is not consecutive. However, all of the pertinent repair information and data for the 'Type A' units is included.

# WORKING PRINCIPLES, MAINTENANCE, AND FAULT FINDING

The overdrive is an additional gear unit between the gearbox and propeller shaft. When in operation it provides a higher overall gear ratio than that given by the final drive crown wheel and pinion.

The primary object of an overdrive is to provide open road cruising at an engine speed lower than it would be in normal top gear. This reduced engine speed gives a considerable reduction in petrol consumption and increase in engine life. Overdrive may also be used on the indirect gears to enhance performance or to provide easy and clutchless gear changing for example in town traffic.

Two basic sizes of unit are produced, known as 'A' and 'D' illustrated in Figs. 1 and 2 respectively. The former is the larger unit and is used on cars having engine capacities of about 2 litres and upwards and the 'D' type on smaller cars.

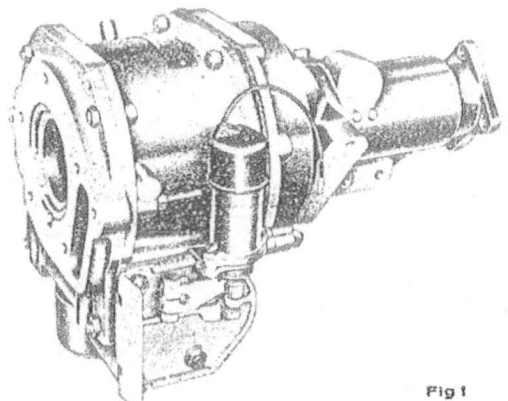
Fig 1

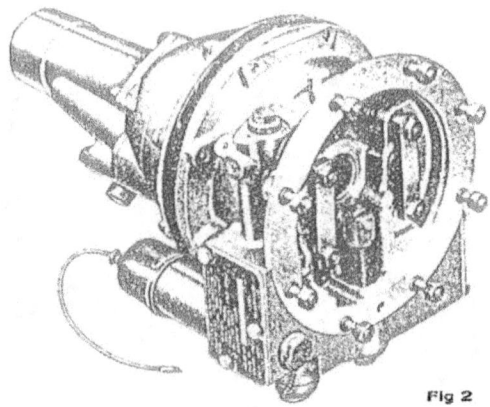
Fig 2

The 'A' type is available with gear ratios of either 0·778 or 0·820 to 1 and the 'D' type with ratios of either 0·756 or 0·802 to 1.

The overdrive is operated by an electric solenoid controlled by a switch, usually mounted on the steering column or fascia panel. An inhibitor switch is invariably fitted in the electrical circuit to prevent engagement of overdrive in reverse and some or all of the indirect gears.

Overdrive can be engaged or disengaged at will at any speed but usually above, say 30 m.p.h. in top gear. It should be operated without using the clutch pedal and at any throttle opening because the unit is designed to be engaged and disengaged when transmitting full power. The only precaution necessary is to avoid disengaging overdrive at too high a road speed, particularly when using it in an indirect gear, since this would cause excessive engine revolutions.

## WORKING PRINCIPLES ('A' and 'D' TYPES)

The overdrive gears are epicyclic and consist of a central sunwheel meshing with three planet gears which in turn mesh with an internally toothed annulus. The planet carrier is attached to the input shaft and the annulus is integral with the output shaft.

The unit is shown diagrammatically in Fig. 3.

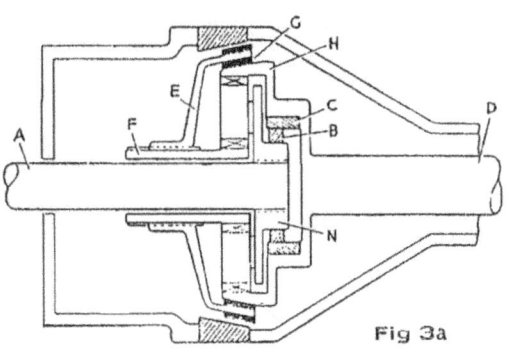
Fig 3a

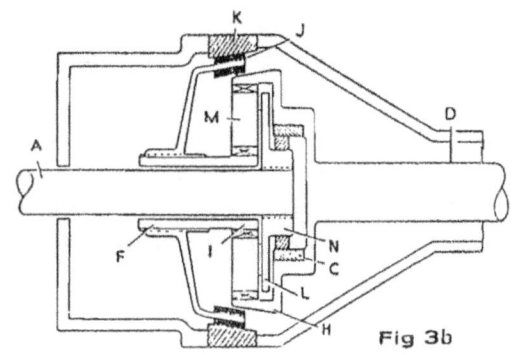
Fig 3b

An extension of the gearbox mainshaft forms the overdrive input shaft. In direct drive (Fig. 3a) power is transmitted from this shaft A to the inner member of a uni-directional clutch N and then to the outer member C of this clutch through rollers B which are driven up inclined faces and wedge between the inner and outer members. The outer member C forms part of the combined annulus H and output shaft D. The gear train is inoperative. A cone clutch E is mounted on the externally splined extension F of the sunwheel and is loaded on to the annulus by a number of springs which have their reaction against the casing of the overdrive unit. The spring load is transmitted to the clutch member through a thrust ring and ball bearing. This arrangement causes the inner friction lining G of the cone clutch to contact the outer cone of the annulus H and rotate with the annulus, whilst the springs and thrust ring remain stationary. Since the sunwheel is splined to the clutch member the whole gear train is locked, permitting over-run and reverse torque to be transmitted. In 'D' type units additional load is imparted to the clutch member, during over-run and reverse, by the sunwheel which, due to the helix angle of its gear teeth, thrusts rearward and has for its reaction member the cone clutch.

Fig. 3b shows the position of the cone clutch when overdrive is engaged. It will be seen that it is no longer in contact with the annulus, but has moved forward so that its outer friction lining J is in contact with a brake ring K forming part of the overdrive casing. The sunwheel I to which the clutch is attached, is therefore held stationary. The planet carrier L rotates with the input shaft A and the planet wheels M are caused to rotate about their own axes and drive the annulus at a faster speed than the input shaft. The uni-directional clutch allows this since the outer member C can over-run the inner member.

Movement of the cone clutch in a forward direction is effected by means of hydraulic pressure which acts upon two pistons when a valve is opened by operating the driver-controlled selector switch. This hydraulic pressure overcomes the springs which load the clutch member on to the annulus and causes the clutch to engage the brake ring with sufficient load to hold the sunwheel at rest.

Hydraulic pressure is developed in the system by a plunger pump, cam operated, from the input shaft. The pump draws oil through a wire mesh filter, in which is incorporated a magnet, and delivers it to the operating valve of the unit. 'A' type units incorporate a hydraulic accumulator in the circuit but in the 'D' type units the working pressure is controlled by a relief valve.

Pressure varies according to the installation but in 'A' type units is usually between 360-520 lbs/sq.ins. and in 'D' type units between 470-520 lbs/sq.ins. (Ferrari 250 Series specification is 490-510 lbs/sq.ins.)

## OPERATION OF OPERATING VALVE 'A' AND 'D' TYPES

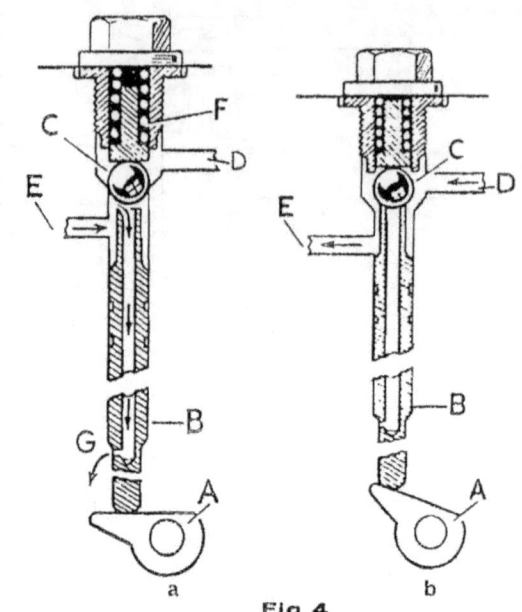

Fig. 4a shows the position of the operating valve in direct drive. In this position the ball C is on the seat in the casing and isolating the supply D from the operating cylinders E. Fig. 4b shows the position of the operating valve in the overdrive position: here the valve has been lifted, by action of the solenoid causing the cam A to rotate, lifting the ball off the seat in the casing and sealing off the top of the valve. This allows oil under pressure to transfer from port D to the operating cylinders E. On returning to direct drive, Fig. 4a the oil from the operating cylinders is exhausted down the hollow stem of the valve and through the restrictor G. On some 'D' type units there is no pressure in direct drive since a port D below the ball seating allows the oil to exhaust to the sump via the hollow operating valve. In overdrive this is sealed off by the ball valve and hence the pressure builds up.

Fig 4

## LUBRICATION

The gearbox and overdrive unit, being adjacent, usually have a common oil supply and the oil level is indicated by a level plug or dipstick in the gearbox. In certain applications the overdrive unit may have an independent supply, in which case a separate filler plug is provided. Separate drain plugs are provided for the gearbox and overdrive unit and both must be removed when draining the oil even though the two systems may be connected. The gauze filter in the overdrive unit should be removed and cleaned whenever the oil is changed. In 'A' type units the filter is accessible when the drain plug is removed. On 'D' type units remove the rectangular plate Ref. A on Fig. 6 which is secured by four setscrews.

Later 'D' type units have lubrication via a drilling in the mainshaft; the spill oil from the relief valve is diverted through drilled passages to a bush in the front casing, and from this into the shaft and along the centre drilling to the rear bearing in the annulus. From here the oil passes due to centrifugal force through the uni-directional clutch to an oil thrower from which it is picked up by a catcher on the planet carrier and to the planet bearings via the hollow planet bearing pins.

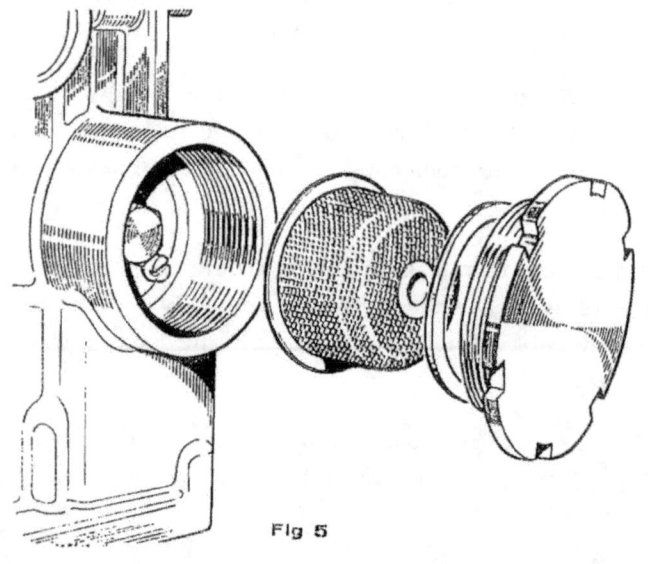

Fig 5

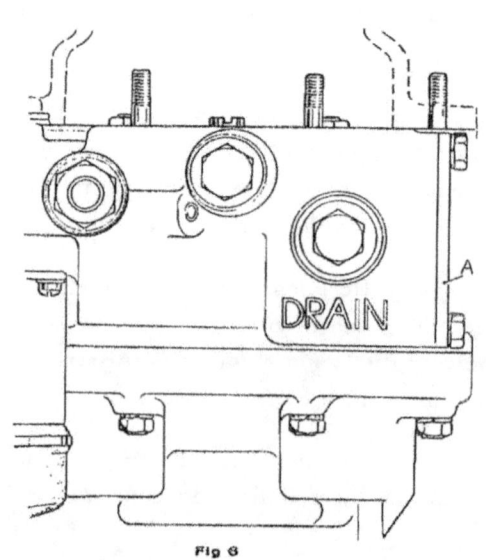

Fig 6

It is essential that an approved lubricant be used when refilling, preferably a straight mineral oil with a viscosity between SAE.30 and SAE.50 with no E.P. additives.

ON NO ACCOUNT SHOULD ANY ANTI-FRICTION ADDITIVES BE PUT INTO THE OIL.

After refilling the gearbox and overdrive, re-check the oil level after the car has been run for a short distance as a certain amount of oil will be distributed round the hydraulic system. It is most important to use clean oil at all times and great care must be taken to avoid the entry of dirt whenever any part of the casing is opened. Dirt, or even lint from a wiping cloth, which finds its way into a valve, will cause trouble. If the hydraulic valves are dismantled, care should be taken to prevent scratches or nicks since these might cause leakage.

FAULT FINDING 'A' and 'D' TYPES

Overdrive does not engage

1. Insufficient oil in gearbox.
2. Electrical system not working. See - The Electrical Circuit.
3. Solenoid operating lever out of adjustment.
4. Insufficient hydraulic pressure due to pump non-return valve incorrectly seating (Probably dirt on seat).
5. Insufficient hydraulic pressure due to worn accumulator on 'A' types, sticking or worn relief valves 'D' types.
6. Pump not working due to choked filter.
7. Pump not working due to damaged pump roller or cam.
8. Leaking operating valve due to dirt on ball seat.
9. Damaged parts within the unit requiring removal and inspection.

Overdrive does not disengage

NOTE   IF OVERDRIVE DOES NOT DISENGAGE DO NOT REVERSE THE CAR OTHERWISE EXTENSIVE DAMAGE MAY RESULT.

1. Fault in electrical control system.
2. Solenoid sticking.
3. Blocked restrictor jet in operating valve.
4. Solenoid operating lever incorrectly adjusted. See - Adjustment of Solenoid Operating Levers.
5. Sticking clutch. See - Sticking Clutch.
6. Damaged gears, bearings, or sliding parts within the unit.

Clutch slip in overdrive

1. Insufficient oil in gearbox.
2. Solenoid lever out of adjustment.
3. Insufficient hydraulic pressure due to pump non-return valve incorrectly seating. (Probably dirt on seat).
4. Insufficient hydraulic pressure due to worn accumulator on 'A' types, sticking or worn relief valve 'D' types.
5. Operating valve incorrectly seated.
6. Worn or glazed clutch lining.

Clutch slip in reverse or free wheel condition on overdrive

1. Solenoid operating lever out of adjustment.
2. Partially blocked restrictor jet in operating valve.
3. Worn or burnt inner clutch lining.

   NOTE   Before removing any of the valve plugs it is essential to operate the solenoid several times in order to release all hydraulic pressure from the system. To do this, engage top gear, switch on the ignition and operate the overdrive control switch several times.

## THE OPERATING VALVE ('A' and 'D' TYPES)

The valve plug is located on top of the unit and access to it is through the floor of the car, a cover plate usually being provided for this purpose. Operate the solenoid several times to release hydraulic pressure from the system. Unscrew the valve plug with a 7/16" A/F spanner. If very tight, a sharp tap on top will help. Remove the spring, plunger and ball. A small magnet will be found useful for this operation. The operating valve can be removed by inserting a piece of stiff wire down its centre and drawing it up, but care must be taken to avoid damaging the seating at the top of the valve. Near the bottom of the valve will be seen a small hole, breaking through to the centre drilling. Fig. 4a - G. This is for the exhaust of oil from the operating cylinders. Ensure that this is not choked.

If necessary the ball can be re-seated on top of the operating valve by placing the ball on a block of wood and sharply tapping the valve after positioning it on the ball. Clean the valve seat in the casing and if necessary re-seat the ball by tapping it gently on its seat with a copper drift. Do not tap the ball too hard or the mouth of the hole will be closed up so that the valve cannot be re-assembled.

## ADJUSTMENT OF SOLENOID OPERATING LEVERS

The operating valve referred to above is lifted by a cam on a transverse shaft. The solenoid operates a lever attached to this shaft. When the solenoid operates the valve must be fully opened.

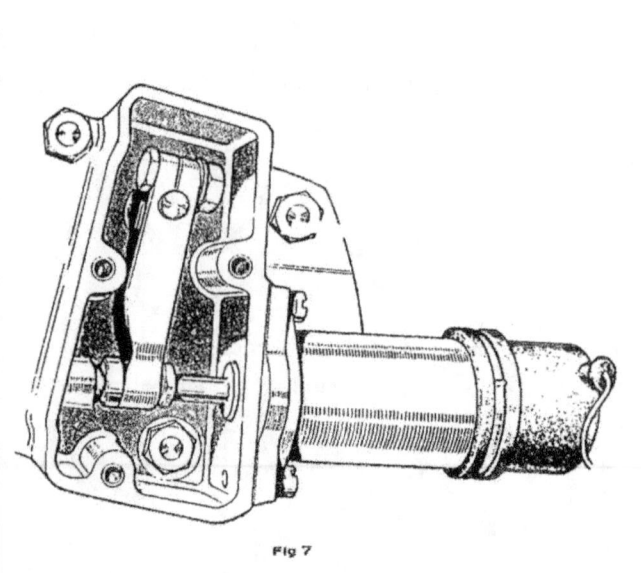

Fig 7

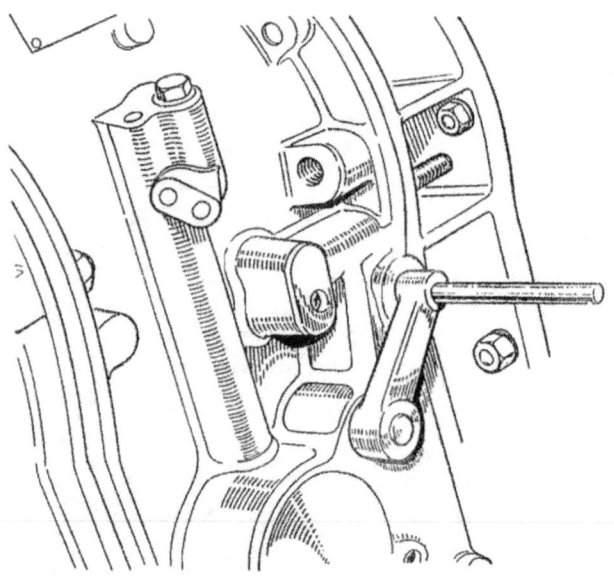

Fig 8

ADJUSTMENT FOR 'A' TYPE UNITS

In this unit the transverse shaft passes right through the casing; where it protrudes there is a setting lever attached. This has a 3/16" hole in its outer end, (Fig. 8). This hole should align with a similar hole in the overdrive casing when the solenoid is energised. For the purpose of checking the setting of the solenoid lever a 3/16" diameter pin, such as a drill shank, should be inserted through the hole in the lever and should register in the hole in the casing when the solenoid is energised. If the pin will not register in the casing the solenoid lever requires adjustment — proceed as follows. Remove the cover plate from the solenoid housing (not fitted on some models). Loosen the clamp bolt on the lever, then rotate the shaft until the pin inserted in the lever, registers in the hole in the casing. Push the solenoid plunger <u>as far home as it will go,</u> and hold the lever fork lightly against the collar on the plunger. Tighten the clamp bolt; remove the pin from the setting lever, then re-check by energising the solenoid and checking the alignment of the holes.

ADJUSTMENT FOR 'D' TYPE UNITS

First remove the rectangular solenoid cover plate which is secured by three screws. Now the solenoid lever can be observed. This also has a 3/16" hole for setting purposes. The procedure is similar to the 'A' type but there is no clamp bolt on the lever. Move the lever until the 3/16" pin pushed through the hole in the lever registers in the hole in the casing, then screw the nut on the plunger until, when the plunger is pushed right home the nut just contacts the forks of the lever. Remove the 3/16" pin. Re-check by energising the solenoid and checking the alignment of the holes. When the solenoid is energised the current consumption should be about 1 ampere. If it is 15-20 amperes it is an indication that the solenoid plunger is not moving far enough to switch from the operating to the holding coil of the solenoid and the lever must be adjusted.

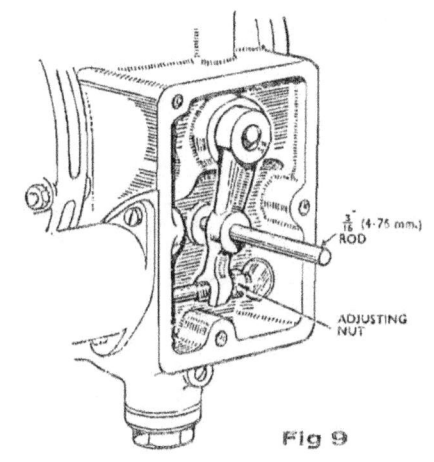

Fig 9

THIS IS IMPORTANT AS HIGH CURRENT WILL CAUSE SOLENOID FAILURE.

TESTING OIL PRESSURE

Release the hydraulic pressure by switching on the ignition, engaging top gear and operating the overdrive switch several times. Remove the operating valve plug and replace it with the hydraulic test equipment (Churchill tool L.188) which has a pressure gauge reading to 800 p.s.i.

Jack up the rear wheels of the car securely, start the engine, engage top gear and run up to about 20 m.p.h. on the speedometer. Hydraulic pressure should then be recorded. Check the pressure in direct and overdrive.

NOTE   On some 'D' type units there is no hydraulic pressure in direct drive but pressure will be recorded when the overdrive is engaged.

Failure to register pressure with overdrive selected may indicate that the pump non-return valve requires cleaning and re-seating.

On those units which normally have pressure in direct drive as well as overdrive, variation in pressure between the two conditions may indicate that the operating valve requires cleaning and re-seating.

THE PUMP VALVE

If the unit fails to operate after re-seating the operating valve, check that the pump is working. Jack up the rear wheels of the car securely, remove the operating valve plug referred to above and start the engine. Engage top gear and with the engine running slowly, watch for oil being pumped into the valve chamber. If none appears the pump is not functioning and its non-return valve should be cleaned. A flow of oil does not necessarily mean that the hydraulic pressure is correct.

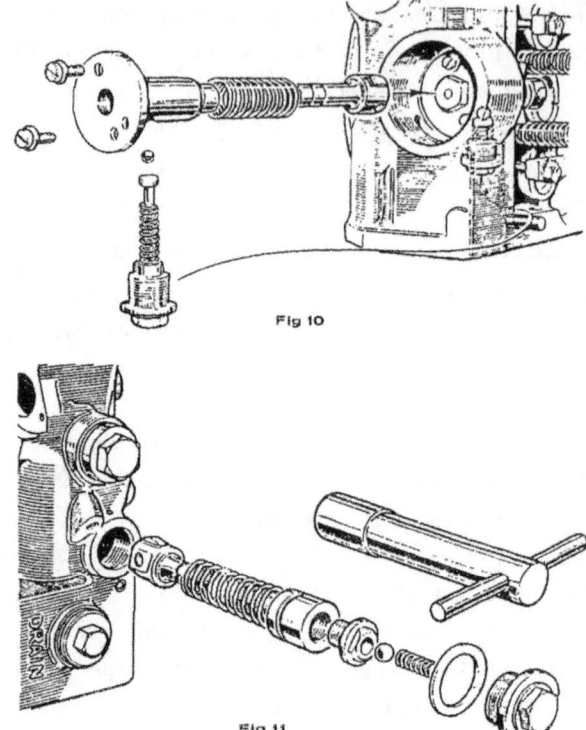

Fig 10

Fig 11

Access to the non-return valve in 'A' type units requires removal of the accumulator end cover. On most units this is also the solenoid bracket. Proceed as follows : Drain off the oil; remove the cover from the solenoid bracket and remove the solenoid. Slacken off the clamp bolt in the solenoid lever and remove the lever and solenoid plunger. Remove the distance collar under the lever. The solenoid bracket is secured by two 5/16" studs and two bolts. Remove the nuts from the studs before unscrewing the bolts; this is important. Now unscrew the bolts together, releasing the compression on the accumulator spring. Remove the spring and guide tube. The pump valve plug will then be seen inside the cavity. The valve consists of a spring, plunger and ball similar to those used for the operating valve, except that the steel ball is $\frac{1}{4}$" diameter. Carefully clean the ball and the valve seating; if necessary re-seat the ball by tapping it sharply on to its seating. When re-assembling, the solenoid lever must be correctly set as already described. See - Adjustment for 'A' Type Units.

The pump valve of 'D' type units is accessible from underneath the unit when the centre plug is removed, Fig. 6. Unscrew the valve body, carefully clean the ball and the valve seating and re-seat the ball by tapping it sharply on to its seating.

STICKING CLUTCH

If overdrive cannot be disengaged after carrying out the procedure outlined previously the trouble is probably caused by a sticking cone clutch. This trouble might be experienced on a new unit due to insufficient "bedding in" of the clutch, but is unlikely to occur on a unit which has been in service for some time.

The clutch can usually be freed by giving the brake ring several sharp blows with a hide mallet. On most cars this can be done from underneath when the car is on a hoist. On some cars, where the gearbox cover is removable, it can be done from above.

THE ELECTRICAL CIRCUIT

Before embarking on the full procedure for fault location, it will be found helpful to keep the following points in mind.

Many operational failures are due to corroded terminals and faulty wiring, so make a point of checking over the wiring and connections first.

96

Good earth connections are essential on all earthed components. This applies particularly to the solenoid because of the heavy current passed momentarily each time the overdrive is engaged.

Incorrect adjustment of the solenoid, resulting in failure of the main winding contacts to open, may cause damage to the solenoid and to the relay.

The method of controlling the overdrive unit differs according to the requirements of the car manufacturer, but the basic system is illustrated in Fig. 12. If overdrive fails to operate and the wiring has been checked, proceed as follows :-

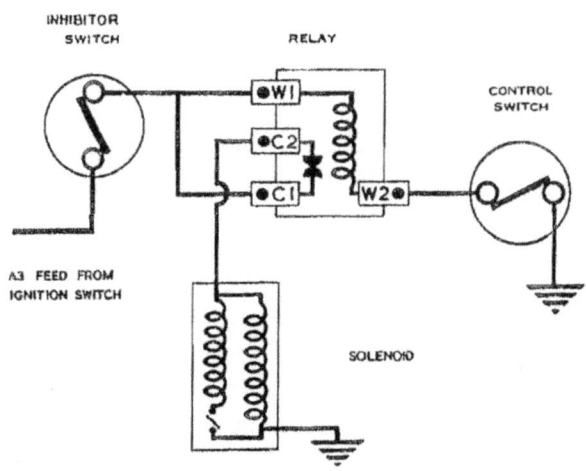

Fig 12

1. Short out terminals of C1 and C2 of the relay, switch on ignition and engage top gear. If the solenoid fails to operate suspect faulty inhibitor switch. If the solenoid operates the inhibitor switch and solenoid are in order. Proceed to Test 4.

2. Connect terminal C1 to A3. If solenoid operates the inhibitor switch is faulty. If it does not operate suspect faulty solenoid.

3. Connect solenoid terminal to A3. If solenoid fails to operate or is sluggish, it is faulty.

4. Connect W1 of relay to A3 with the control switch closed. If solenoid operates, relay and control switch are satisfactory. If solenoid does not operate proceed to Test 5.

5. Link W2 of relay to earth. If solenoid fails to operate the relay is faulty. If solenoid operates, control switch may be faulty. Proceed to Test 6.

6. Connect feed terminal of control switch to earth with switch closed. If solenoid operates, control switch is faulty.

# KEY TO FIGURE 20

1. Clutch Thrust Ring Assy.
2. Clutch Return Springs
3. Clutch Sliding Member
4. Thrust Ballrace
5. Circlip
6. Circlip
7. Brake Ring
8. Sunwheel Assy.
9. Planet Carrier Assy.
10. Annulus Assy.
11. Bronze Thrust Washer
12. Bronze Thrust Washer
13. Steel Thrust Washer
14. Cage (Uni-Directional Clutch)
15. Rollers (Uni-Directional Clutch)
16. Spring (Uni-Directional Clutch)
17. Inner Member (Uni-Directional Clutch)
18. Thrust Washer
19. Annulus Front Ballrace
20. Annulus Rear Ballrace
21. Selective Spacing Washer
22. Rear Casing
23. Studs
24. Speedometer Pinion
25. Speedometer Pinion Pilot Bush
26. Speedometer Pinion Support Bush
27. Dowel Screw
28. Copper Washer
29. Rear Oil Seal
30. Coupling Flange
31. Coupling Flange Nut
32. Plain Washer
33. Overdrive Joint Washer
34. Pump Operating Cam
35. Bridge Piece
36. Operating Pistons
37. Sealing Ring
38. Operating Valve
39. Breather
40. Operating Valve Plug
41. Operating Valve Spring
42. Operating Valve Spring Plunger
43. Operating Valve Ball
44. Support Bushes
45. Main Casing
46. Guide Peg
47. Pump Plunger
48. Pump Roller
49. Pump Roller Pin
50. Pump Plunger Spring
51. Pump Body
52. Pump Body Screws
53. Pump Body Base Plug
54. Oil Filter
55. Sealing Washer
56. Drain Plug
57. Non Return Valve Ball
58. Non Return Valve Plunger
59. Non Return Valve Spring
60. Non Return Valve Plug
61. Valve Setting Lever
62. 'O' Ring
63. Cam Lever
64. Operating Lever Shaft
65. Solenoid Bracket Joint
66. Solenoid Bracket
67. Rubber Stop
68. Distance Collar
69. Operating Lever
70. Solenoid Cover Joint
71. Solenoid Cover
72. Solenoid
73. Sealing Disc
74. Accumulator Sleeve
75. 'O' Ring
76. Piston Rings
77. Accumulator Piston
78. Accumulator Spring
79. Accumulator Tube

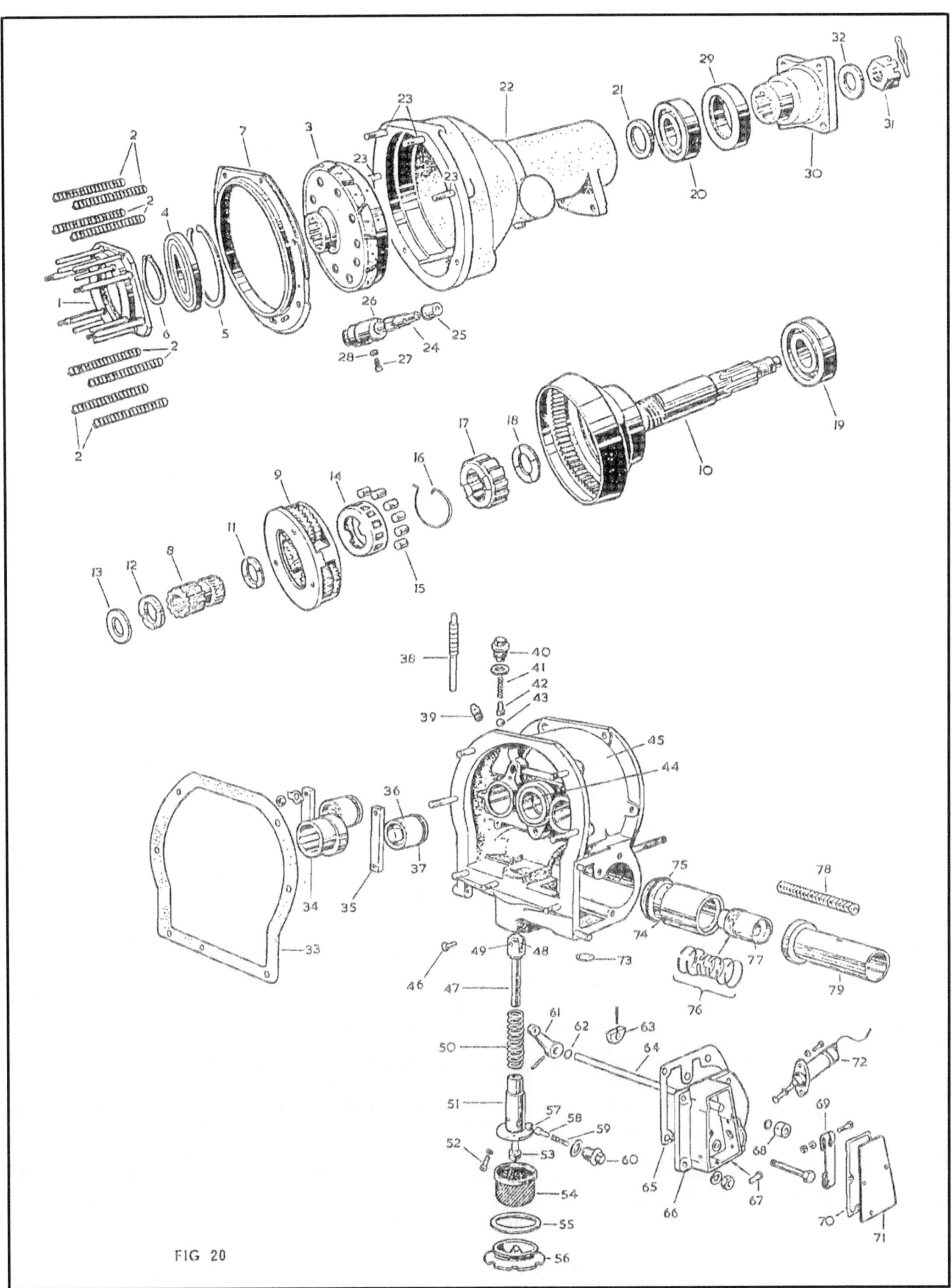

FIG 20

## OVERDRIVE REMOVAL

The overdrive unit may be removed without taking the gearbox from the car providing that sufficient clearance 7.0" (177 mm) exists to allow the overdrive to be moved rearwards. However, it is usually more convenient to remove the gearbox and overdrive complete. Whichever means is adopted proceed as follows:-

The unit is split at the rear face of the adapter casing. It will be seen that there are four or five short studs and two long studs (or in some cases bolts). Remove the nuts from the short studs first, then simultaneously loosen the nuts on the long studs, thus releasing the pressure on the clutch return springs. Pay attention to the degree of stiffness which is given to these two nuts by the pressure of the clutch springs, so that extra pressure required to re-tighten these nuts later on will be anticipated. Remove the two nuts, then the overdrive unit can be withdrawn off the mainshaft.

The overdrive can be divided into four main assemblies :-

1. FRONT CASING AND BRAKE RING
2. CLUTCH SLIDING MEMBER
3. PLANET CARRIER AND GEAR TRAIN
4. REAR CASING AND ANNULUS

## DISMANTLING

**IMPORTANT**

SCRUPULOUS CLEANLINESS MUST BE MAINTAINED THROUGHOUT ALL SERVICE OPERATIONS. EVEN MINUTE PARTICLES OF DUST OR DIRT, OR LINT FROM CLEANING CLOTHS MAY CAUSE DAMAGE, OR AT BEST INTERFERE WITH CORRECT OPERATION.

Prepare a clean area in which to lay out the dismantled unit, and some clean containers to receive the small parts.

In installations where the oil supply is common with the gearbox, it follows that the same high standards of cleanliness must be maintained, when servicing the gearbox.

SPECIAL TOOLS

A complete list of special tools can be obtained, as listed in the Appendix.

For the initial examination, dismantle into the four main assemblies proceeding as follows :-

Hold the overdrive with front casing uppermost in a vice fitted with suitable soft jaws.

Remove all the clutch return springs from their pins, noting that the four springs nearest to the centre of the unit are shorter than the outside springs.

Release the tabwashers locking the four $\frac{1}{4}$" nuts, retaining the operating piston bridge pieces. Remove the nuts, tabwashers and bridge pieces.

Remove the six nuts which secure the front and rear casings. (On some units it may be necessary to remove the solenoid in order to gain access to one of the nuts). Separate the two casings. The brake ring is spigoted into each half and may remain attached to the front half, if not a few taps with a mallet around its flange will remove the brake ring from the rear casing.

Remove one steel and one bronze thrust washer from the forward end of the sunwheel, noting their positions.

Lift out the clutch sliding member complete with the thrust ring and bearing.

Lift out the sunwheel and the bronze washer situated in the recess in the planet carrier.

NOTE  In the case of the 22% unit, it is not possible to remove this washer because the planet gears overlap it. Lift out the planet carrier assembly.

The overdrive is now divided into the four main assemblies.

FRONT CASING AND BRAKE RING

If on road test the hydraulic system was performing correctly and there has been no major failure causing metal dust etc., it should not be necessary to dismantle this assembly any further, but if it is, proceed as follows :-

Remove the operating valve plug, lift out the spring, plunger and ball, remove the operating valve by lifting from underneath and grasping it as it protrudes from the casing. Place all the components in a clean container, taking care not to damage the valve seating.

Remove the operating pistons by gripping the centre bosses with a pair of pliers and applying a rotary pull.

The Solenoid

To remove the solenoid, first take off the solenoid cover plate (when fitted). Remove the two screws and pull off the solenoid, ease the plunger out of the yoke of the valve operating lever.

Release the clamp bolt on the valve operating lever; remove the lever and the collar under it from the valve operating shaft.

The Accumulator

Access to the accumulator is gained by removing the solenoid bracket as follows. First remove the nuts from the studs then simultaneously loosen the two setscrews painted red, securing the bracket to the casing.

The setscrews are of sufficient length to allow the accumulator spring to be completely released, AND SHOULD ALWAYS BE REMOVED AFTER THE NUTS.

After removing the bracket, the accumulator spring is exposed.

There are three alternative sizes of accumulator piston, namely, $1\frac{1}{8}$" diameter, $1\frac{1}{2}$" diameter and $1\frac{3}{4}$" diameter.

The $1\frac{1}{8}$" diameter has one spring and a tube.

The $1\frac{1}{2}$" diameter has two springs and a tube.

The $1\frac{3}{4}$" diameter has two springs and no tube.

Remove the springs and tube, whichever is applicable.

The accumulator sleeves for the $1\frac{1}{8}$" diameter and $1\frac{1}{2}$" diameter piston are removed with the aid of special tools L 182 and L 216 respectively. Insert the special tool into the accumulator sleeve and tighten the lower wing nut. Withdraw the accumulator sleeve and piston complete by applying a rotary pull to the upper wing bolt of the tool Fig. 21. Place the assembly into a clean container to avoid soiling or damaging the rubber sealing rings.

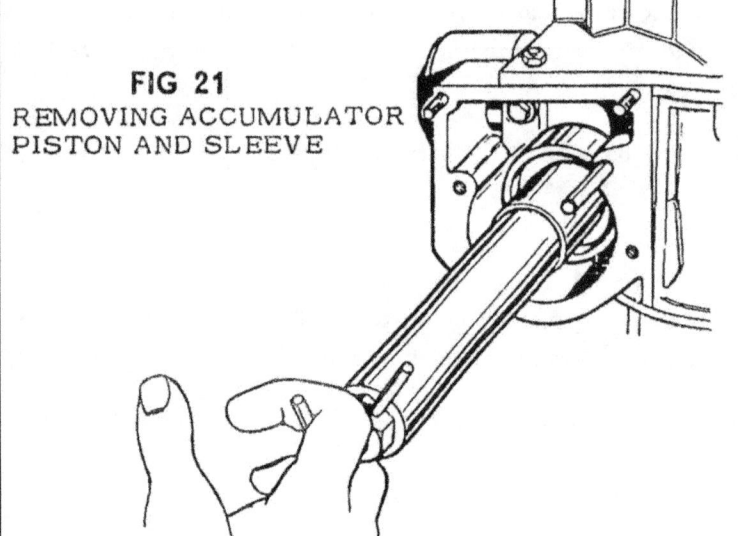

**FIG 21**
REMOVING ACCUMULATOR PISTON AND SLEEVE

The $1\frac{3}{4}$" diameter piston is removed by screwing a $\frac{3}{8}$" UNF screwed rod into the piston and withdrawing it with a rotary pull.

The Pump Non-Return Valve

This valve is located in the cavity exposed by removing the solenoid bracket and is adjacent to the accumulator bore. Remove the hexagon plug Ref. 60, and lift out the spring plunger and $\frac{1}{4}$" diameter ball.

The Filter

Remove the brass drain plug, Ref. 56, lift out the filter Ref. 54. Located in the recess of the plug, are 3 magnetic plastic rings.

---

IMPORTANT

IT IS IMPERATIVE THAT THE PUMP NON-RETURN VALVE IS REMOVED BEFORE ATTEMPTING TO REMOVE THE PUMP

---

Pump

To remove the pump, remove the filter and pump non-return valve as described previously. Remove the two pump retaining screws Ref. 52 and the base plug Ref. 53. The pump body can now be extracted, using Tool No. L 183 as follows :-

Screw the short threaded portion of the spindle into the pump body from where the base plug was removed then place the adapter in position against the casing and screw up the wing nut, thereby pulling the body out of the casing: the plunger and spring will then be removed during this process Fig. 22.

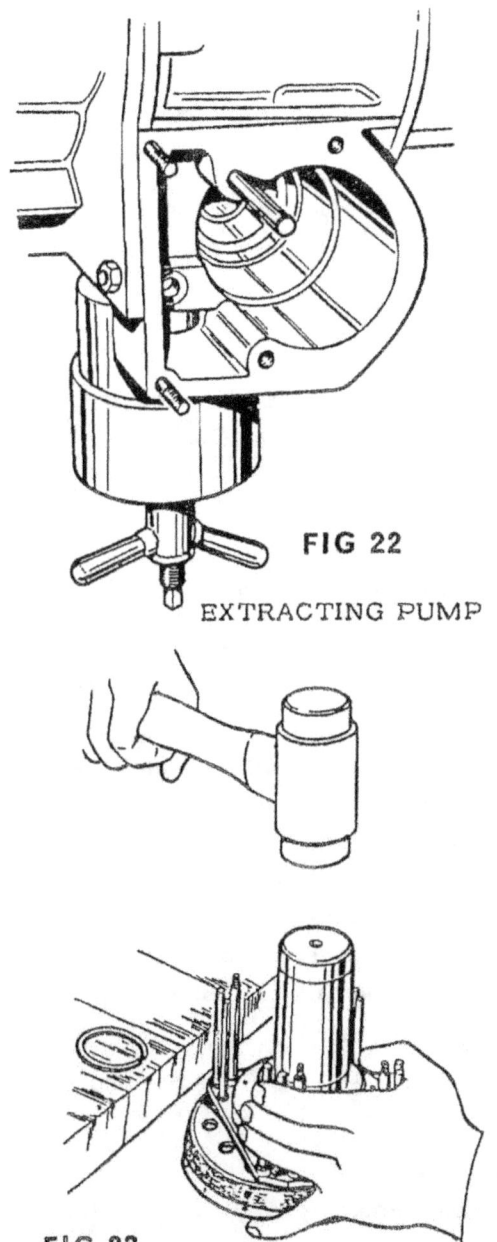

FIG 22
EXTRACTING PUMP

FIG 23
DISMANTLING CONE CLUTCH AND THRUST RING ASSEMBLY

CLUTCH SLIDING MEMBER

Remove the thrust ring complete with bearing from the sliding cone clutch member by withdrawing the circlip from its groove in the forward end of the clutch hub and pressing out the clutch member. Care must be taken not to distort the clutch member or damage the linings.

Remove the thrust bearing Ref. 4 from the thrust ring by removing the large circlip Ref. 5 and pressing out the bearing Fig. 23.

PLANET CARRIER ASSEMBLY

At this stage inspect all the gear teeth for any signs of damage or chipping, and assess the fit of the assembled bearing for any excessive clearance.

For models where replacement planet gears are not available separately for servicing, a complete planet carrier sub assembly should be substituted if damage or wear necessitates replacement.

Replacement planet shafts and bearings (except for caged type) are, however, available for all models.

NOTE  Caged bearings can only be supplied together with new gears.

In cases where planet gears are available separately, they must be installed in sets of three, even though only one or two of the original planet gears were damaged.

To extract the pins proceed as follows :—

IMPORTANT   Remove one gear at a time and mark by scribing the individual gear, planet pin and relative planet hole location in the carrier to ensure that each gear is refitted into its original location.

NOTE   Each gear is marked with a dot by the manufacturer. This is for angular relationship in assembly of the compound gears and is also used to obtain the correct angular position of the Planet Carrier and Gear Train before assembly with the Annulus. See section on re-assembly of the Planet Carrier and Gear Train.

Support the planet carrier on a suitable hollow abutment through which the pin will pass. Using a drift, drive out the pin, shearing the small Mills pin which secures it. Knock or drill out the broken end of the Mills pin from the carrier and planet pin.

Some units have an oil catcher rolled on to the rear face of the planet carrier making it impossible to drive the planet pins out in the aforementioned manner. In this case a new planet carrier assembly complete must be fitted.

<u>To Extract the Needle Bearings (Using Tool No. L 203)</u>

Secure the square ended shank of the tool vertically in a vice and remove the wing nut and all the collars. Slide the gear over the spindle and allow the bearing to abut against the spindle shoulder. Fit the main body and wing nut, and press the gear off the bearings.

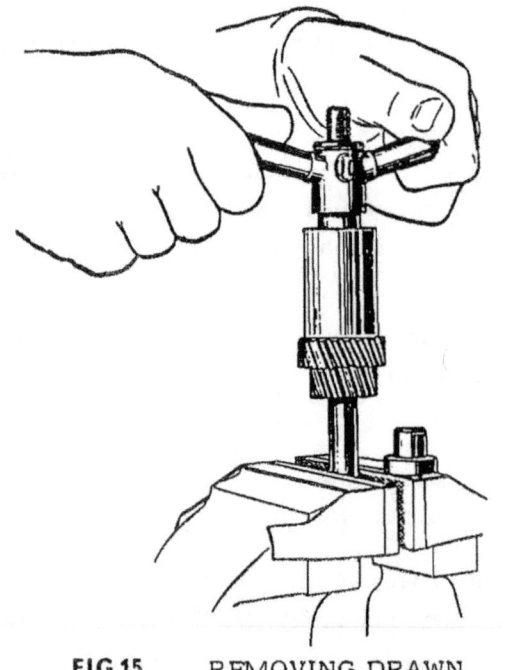

**FIG 15**   REMOVING DRAWN CUP NEEDLE BEARINGS

REAR CASING AND ANNULUS

To dismantle this assembly proceed as follows :—

Remove the uni-directional clutch Ref. 14 to 17 by placing the special assembly ring (Tool No. L 178) centrally over the front face of the annulus and lifting the inner member of the uni-directional clutch up into it. This will ensure that the rollers do not fall out of the retaining cage. Place the parts in a suitable container. Alternatively, if dismantling further, remove the assembly ring and allow the rollers to come out, and the hub will readily come from the cage, exposing the spring.

Remove the bronze thrust washer fitted between the hub of the uni-directional clutch and the annulus.

<u>Removal of Annulus</u>

Remove speedometer dowel screw Ref. 27, then, using Tool No. L214 to prevent damaging the thread, withdraw the speedometer drive bush and pinion Refs. 26 and 24. Remove the coupling flange Ref. 30.

Remove oil seal (if necessary) by screwing the taper thread of the outer member of the special tool (L176) into it and tightening the centre bolt against the rear of the tail shaft. Press annulus forward out of the rear casing. The front bearing should remain assembled to the annulus, leaving the rear bearing in the casing. Remove the distance collar from its shoulder in front of the splines. Withdraw the front bearing from the annulus, using Tool No. L167 in conjunction with handpress No. RG4221B (See Fig. 24). Drive out the rear bearing from the rear casing.

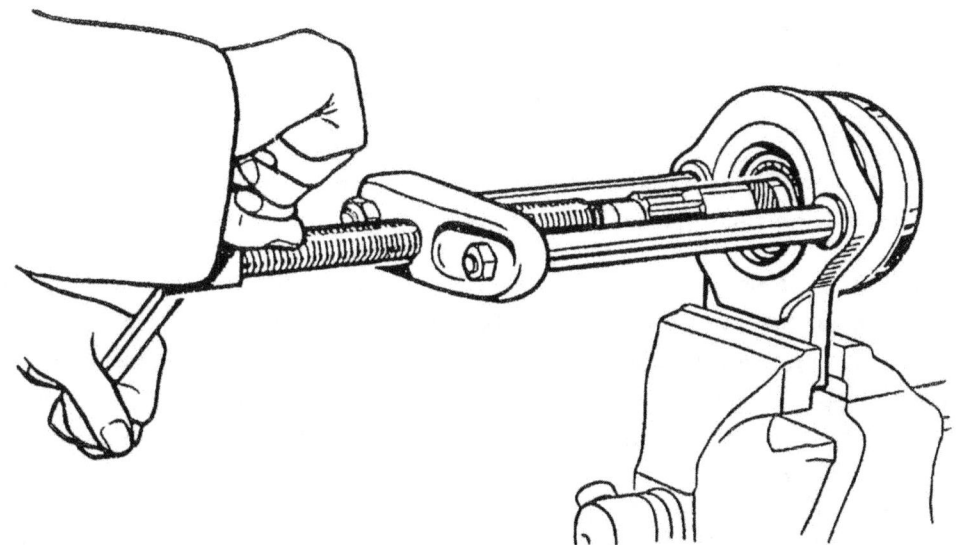

**FIG 24**  REMOVING FRONT BEARING FROM ANNULUS

## INSPECTION

EACH PART SHOULD BE THOROUGHLY CLEANED
AND EXAMINED AFTER THE UNIT IS DISMANTLED

### FRONT CASING AND BRAKE RING

Inspect the front casing for cracks, damage etc. Examine the bores of the operating cylinders and accumulator for scores or wear.

Check for signs of leaks from the plugged ends of the oil passages. Ensure that the sealing disc beneath the accumulator is tight and not leaking. Inspect the centre bore of the support bushes Ref. 44 for wear and damage. Inspect the bronze and steel thrust washers Ref. 12 - 13.

Check operating pistons Ref. 36 for signs of scores and replace sealing rings Ref. 37 using tool No. L 180 if there is any sign of damage or distortion.

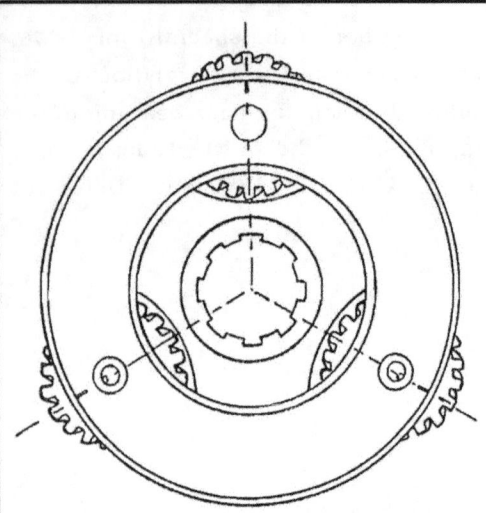

FIG 34 PLANET CARRIER AND GEAR TRAIN SHOWING ETCHED LINES

into the correct angular position before re assembly to the annulus, in accordance with the following procedure. Turn each gear respectively until a dot marked on one tooth of the large gear is positioned radially outwards, Fig. 34. Fit the bronze washer Ref. 11 in the recess in the planet carrier. Insert the sunwheel meshing with the planet gears and keeping the dots in the same position, insert this assembly, meshing the gears in the annulus.

Insert the dummy mainshaft tool No. L 185A at this stage, turning the sunwheel until the shaft engages in both the planet carrier and the uni-directional clutch splines. If any new parts have been fitted in connection with the gear or casings, it becomes necessary to check the end float of the sunwheel which should be between .008" and .014" (.20 - .35 mm).

To do this proceed as follows :- Fit an extra thrust washer of known thickness on top of the sunwheel, over the dummy shaft, then fit the original bronze and steel thrust washers in that order.

Fit the brake ring to the front casing and tap fully home. Fit the front casing over the dummy mainshaft and offer it up to the rear casing. Due to the extra thrust washer mentioned above, the two casings will not meet fully at their flanges. Measure this gap which will represent the thickness of the extra thrust washer, minus the end float of the sunwheel. If the indicated float is more or less than that required it must be adjusted by replacing the steel thrust washer at the front of the sunwheel by one of less or greater thickness as required.

When the correct thickness washer has been ascertained, remove the front casing and the thrust washers and continue with the main assembly.

CLUTCH SLIDING MEMBER

Re-assemble as follows :- Press the thrust bearing evenly into the thrust ring and fit the large circlip.

Press this assembly on to the hub of the clutch sliding member taking great care not to damage the linings and fit the smaller circlip to the clutch sliding member.

Fit this assembly over the sunwheel splines and engage the inner linings on to the annulus.

Fit the bronze washer on the top of the sunwheel, also the steel selective washer of the correct thickness as previously determined.

Smear liquid jointing compound on both sides of the brake ring flange, and tap this home on the front casing.

Fit the front casing and the brake ring to the rear casing, carefully positioning the Thrust Ring pins, through the four holes in the front casing. Fit and tighten nuts on to the six studs.

> NOTE  If the unit has a vertically mounted solenoid, a thin nut is fitted to the stud adjacent to the solenoid cap to provide clearance.

Fit the operating piston bridge pieces Ref. 35 using new tabwashers and nuts.

Fit the distance collar Ref. 68 to the operating lever shaft.

Fit the operating lever Ref. 69. Insert the solenoid plunger into the yoke of the valve setting lever; fit the solenoid with the new joint and tighten the two screws.

Adjust the solenoid operating lever as previously described.

Fit the solenoid cover plate (if applicable) and tighten the appropriate screws, ensuring that the joint washer is in good condition.

The overdrive is now complete and ready for fitting to the gearbox.

Inspect the gearbox mainshaft for nicks and burrs. See that all the oil holes are open and clean.

Check the oil pump operating cam for any undue wear.

Fit the cam to the gearbox mainshaft with the long plain end of the cam towards the gearbox.

If the gearbox has been removed from the car, adopt the following procedure :-

Hold the overdrive vertically in a vice with the front casing uppermost.

Fit the clutch return springs to the respective pins on the thrust ring i.e. the longer springs to the outer pegs. <u>This is most important or the springs will become coilbound thus preventing the correct operation of the overdrive.</u>

Remove the dummy mainshaft; the splines will then be correctly lined up.

Fit a new joint to the front face of the overdrive.

Engage top gear, stand gearbox on end, and enter the mainshaft into the overdrive unit. Turn the primary shaft in the gearbox until the splines engage and then turn further until the lowest portion of the cam coincides with the oil pump roller. Position the clutch springs on the respective bosses on the gearbox rear extension. Press the gearbox down to test the cushioning of the springs.

Fit two nuts to the long studs and tighten, evenly compressing the springs, until there is a gap of approximately $\frac{3}{4}$" between the overdrive casing and the gearbox rear extension, meanwhile <u>ensuring that the oil pump cam does not drop off the splines or the key fall from the mainshaft.</u>

Enter two screw drivers into the gap between the overdrive casing and the gearbox rear extension, with one, compress the oil pump plunger spring and with the other, lever the cam down into alignment with the plunger roller.

Continue tightening the two nuts on the long studs until the faces meet. If the faces fail to meet by about $\frac{5}{8}$" and the nuts become tight, misalignment of the splines is indicated, in which case remove the gearbox from the overdrive again and re-align the splines by rotating the inner member of the uni-directional clutch in an anti-clockwise direction; this can be done by probing with a long screwdriver. Re-check by inserting the dummy mainshaft again.

Re-fit the gearbox to the overdrive following the above procedure.

If the gearbox has been left in the vehicle, the method of fitment remains the same, but particular care must be taken to ensure that the clutch springs are correctly located.

# APPENDIX

## SPECIAL TOOLS FOR 'A' TYPE OVERDRIVE

| Tool No. | Description |
|---|---|
| L 176 A | Drive shaft oil seal remover adapters. (used with Main Tool 7657) |
| L 177 A | Drive shaft oil seal replacer |
| L 178 | Assembly ring for uni-directional clutch |
| L 179 | Piston ring fitting tool $1\frac{1}{8}$" diameter |
| L 180 | Piston ring fitting tool $1\frac{3}{8}$" diameter |
| L 181 | Accumulator O ring replacer |
| L 182 | Accumulator Piston Housing remover |
| L 183 A | Oil Pump Body remover (Main Tool) |
| L 183 A-1 | Oil Pump Body remover adapter |
| L 183 A-2 | Oil Pump Body remover adapter |
| L 184 | Pump Barrel replacer |
| L 185 A | Dummy drive shaft |
| L 186 | Mainshaft Bearing replacer |
| L 187 | Annulus and Tailshaft Bearing remover and replacer - Adapters (used with hand press RG 4221 B) |
| L 188 | Hydraulic Test Equipment (pressure gauge) |
| L 190 A | Tailshaft End Float gauge |
| L 203 | Planet Gear Needle Bearing remover and replacer |

## 250GT ~ BRAKES

| | |
|---|---|
| GENERAL DATA | 110 |
| DRUM BRAKES | 111 |
| DISC BRAKES | 112 - 113 |
| BRAKE SERVO | 114 |

## BRAKE SYSTEM

| MILEAGE | MAINTENANCE |
|---|---|
| Daily | Check pedal pressure before driving at speed. |
| Every 500 Miles | Check brake fluid, top up if necessary. (Use new fluid) |
| Every 1,000 Miles | Check and adjust parking brake. |
| Every 1,500 Miles | Bleed entire brake system, starting from the boosters then each wheel. Never reuse old brake fluid. |
| Every 5,000 Miles | Clean disc rotors, sand paper any surface rust or glaze with #400 emery. |
| Every 5,000 Miles | Change all brake fluid, replacing with approved racing disc fluid from new sealed cans. |
| Every 7,500 Miles | Inspect rubber brake hoses. |
| Every 10,000 Miles | Replace all disc pads. |
| 20,000 Miles | Rebuild disc actuating cylinders, and master cylinder. |
| 30,000 Miles | Rebuild vacuum booster(s). |
| 40,000 Miles | Rebuild rotor discs. |
| 50,000 Miles | Replace all flexible brake hoses. |

### DISC BRAKE PADS

| MODEL | PAD MANUFACTURER | PAD NO. FRONT | PAD NO. REAR |
|---|---|---|---|
| 275GT | Mintex | 875/5201 | 875/4138 |
| 250GT | Mintex | 875/5201 | 875/4138 |
| 330GT | Mintex | 875/5201 | 875/5138 |
| 365GTB/4 | Textar | T252 | T252 |

### RECOMMENDED BRAKE FLUID

Dunlop Racing Brake Fluid
Shell Donax-B-SAE 70 R3
Kelsey-Hayes Kelstar Disc Brake Fluid No. 7999
ATE Tipo H
(Brake fluids other than above should be disc type, and exceed SAEJ1703)
NOTE: Always maintain fluid reservoir at least 1/4 full at all times, and never more than 1/4" from top. Never mix different types of fluid in system.

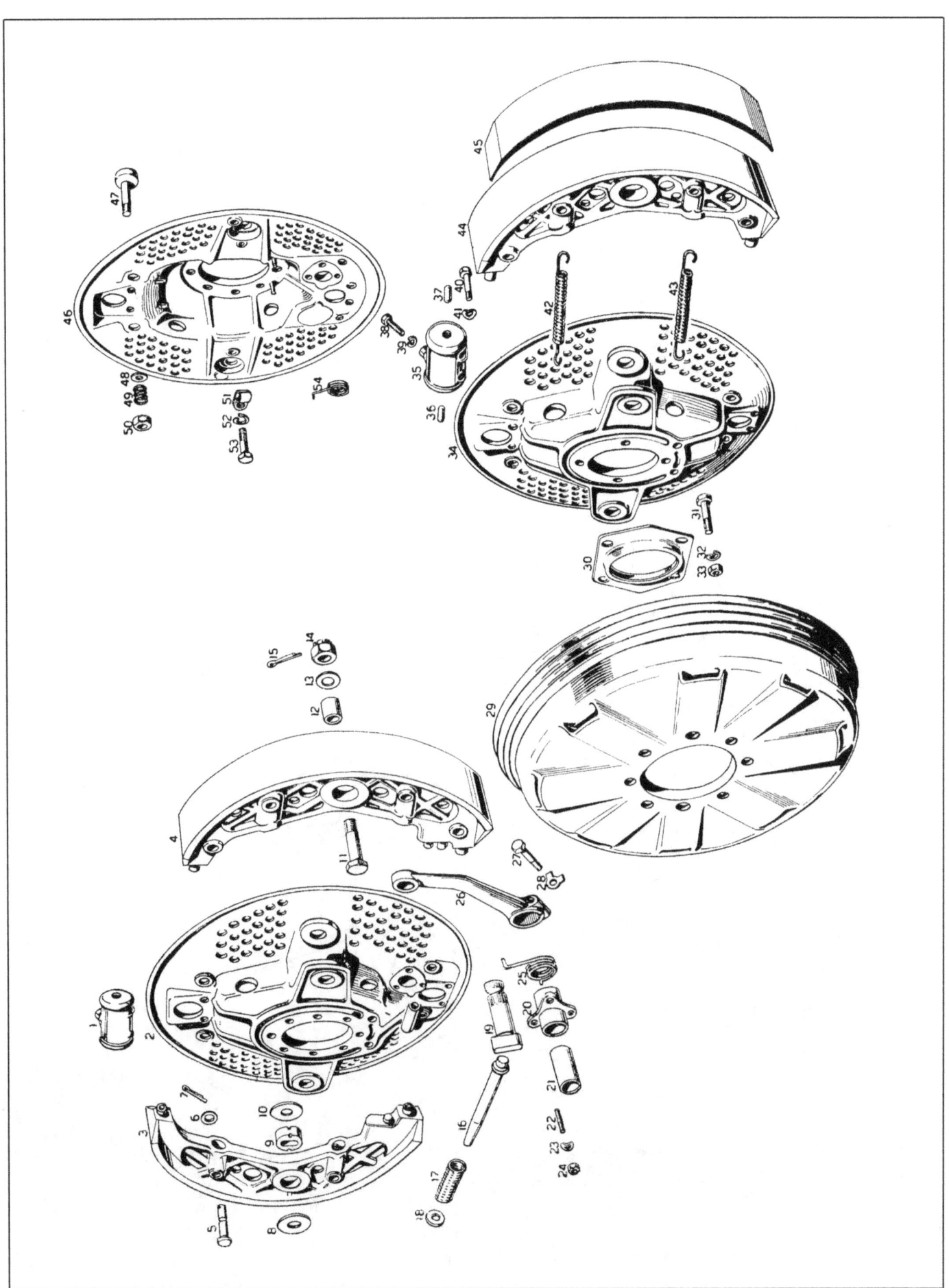

# TAV. 29 - MOZZI - DISCHI FRENO E POMPA DI COMANDO

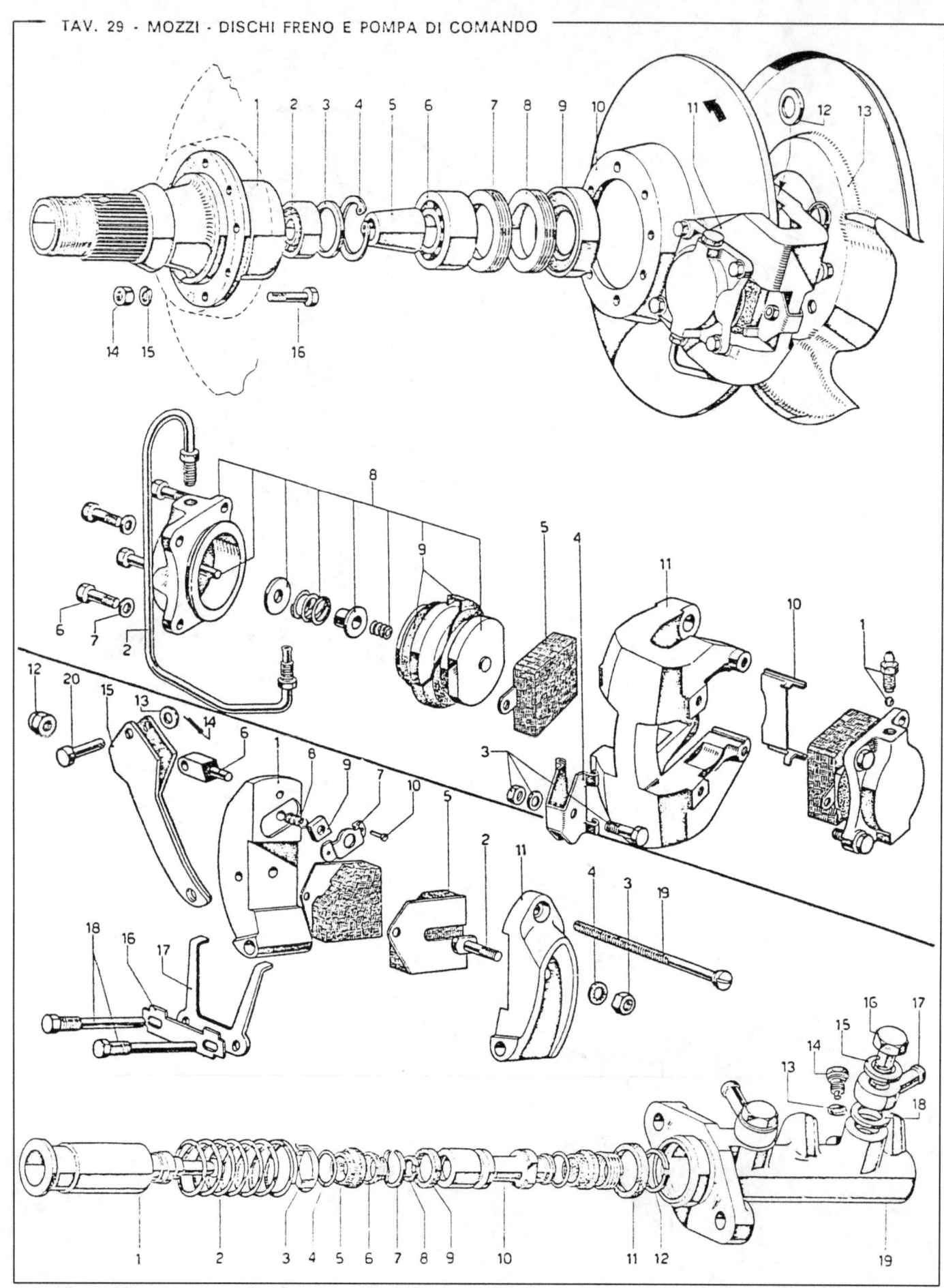

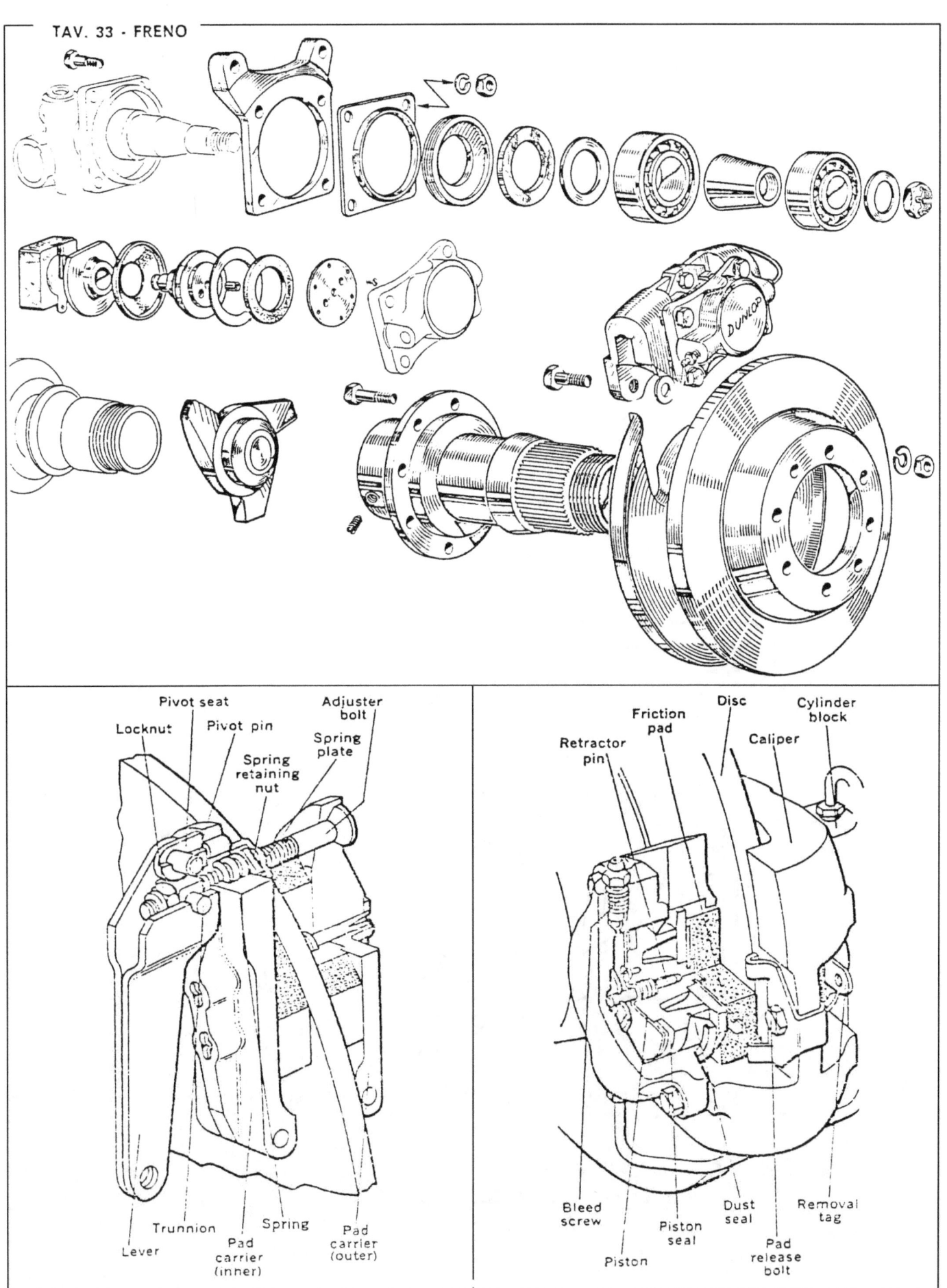

TAV. 33 - FRENO

# Ferrari | Servofreno a decompressione. | Tav. N° 3.

## Servo freno in posizione di riposo.

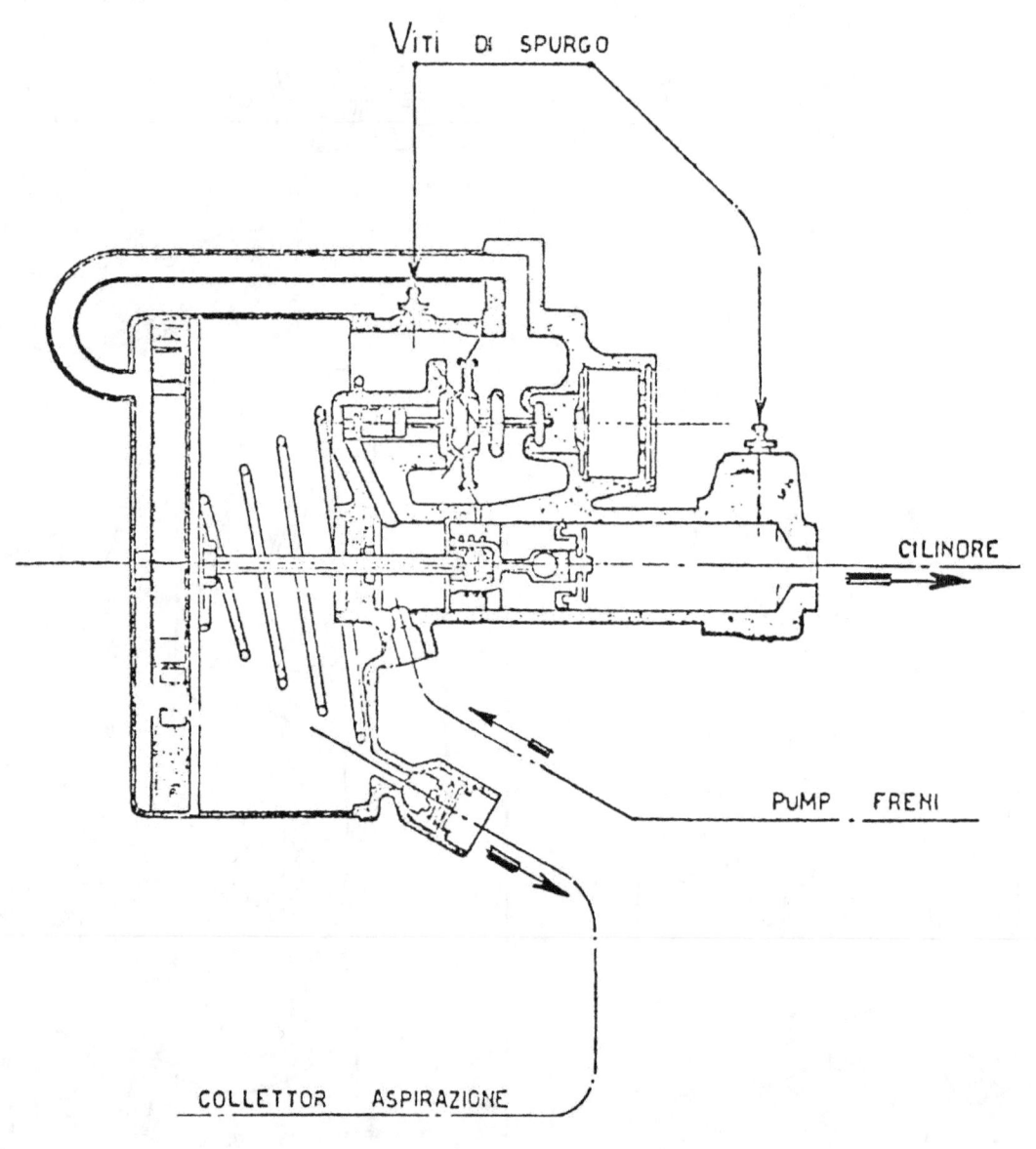

## 250GT ~ REAR AXLE, DRIVESHAFTS, SHOCK ABSORBERS AND STEERING

| | |
|---|---|
| HALFSHAFT - ILLUSTRATION | 116 |
| DRIVE SHAFT - ILLUSTRATION | 117 |
| DIFFERENTIAL - ILLUSTRATION | 118 - 119 |
| SHOCK ABSORBERS - GENERAL DATA | 120 - 121 |
| STEERING BOX - ILLUSTRATION | 122 |

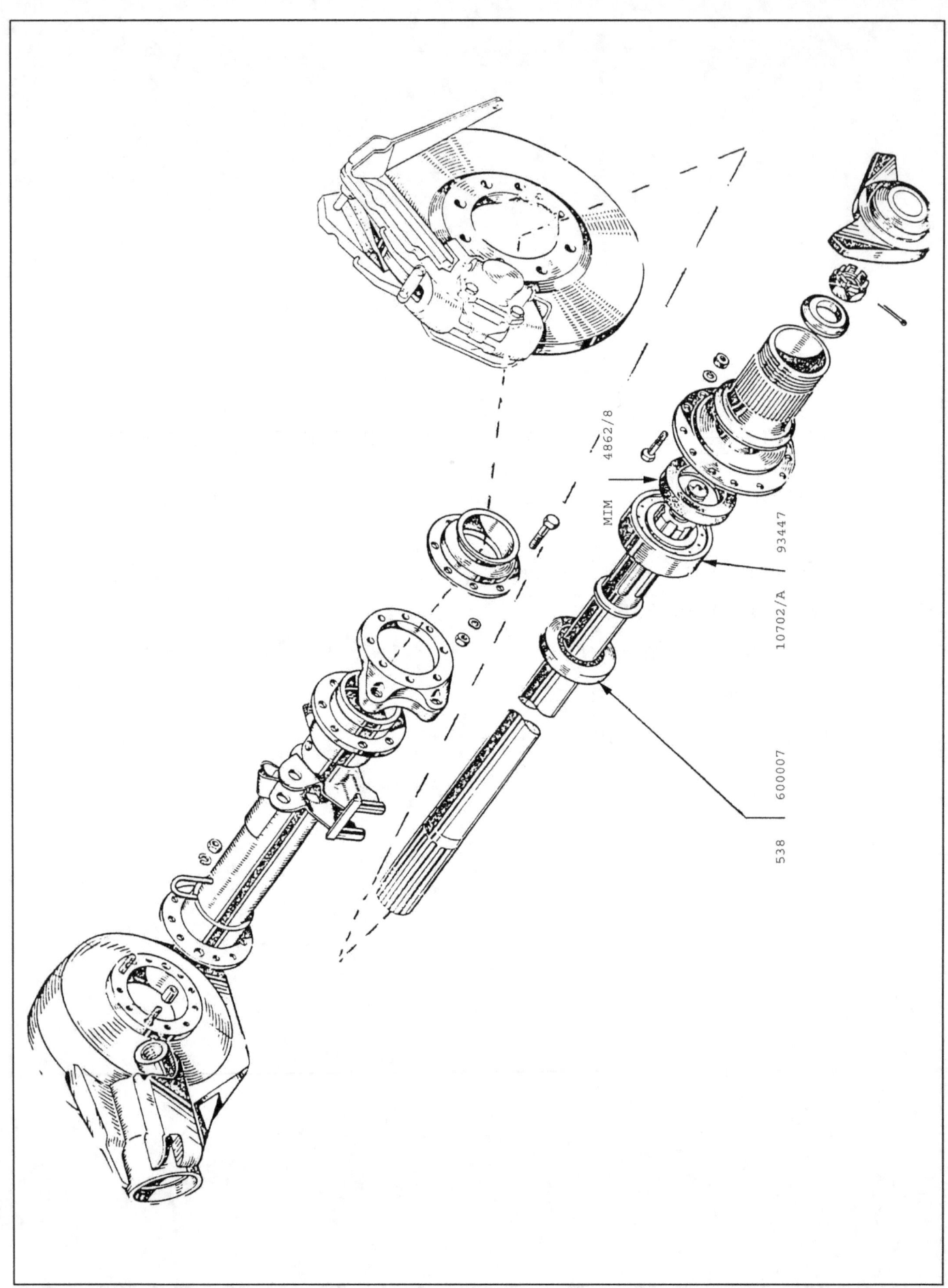

ALBERO DI TRASMISSIONE

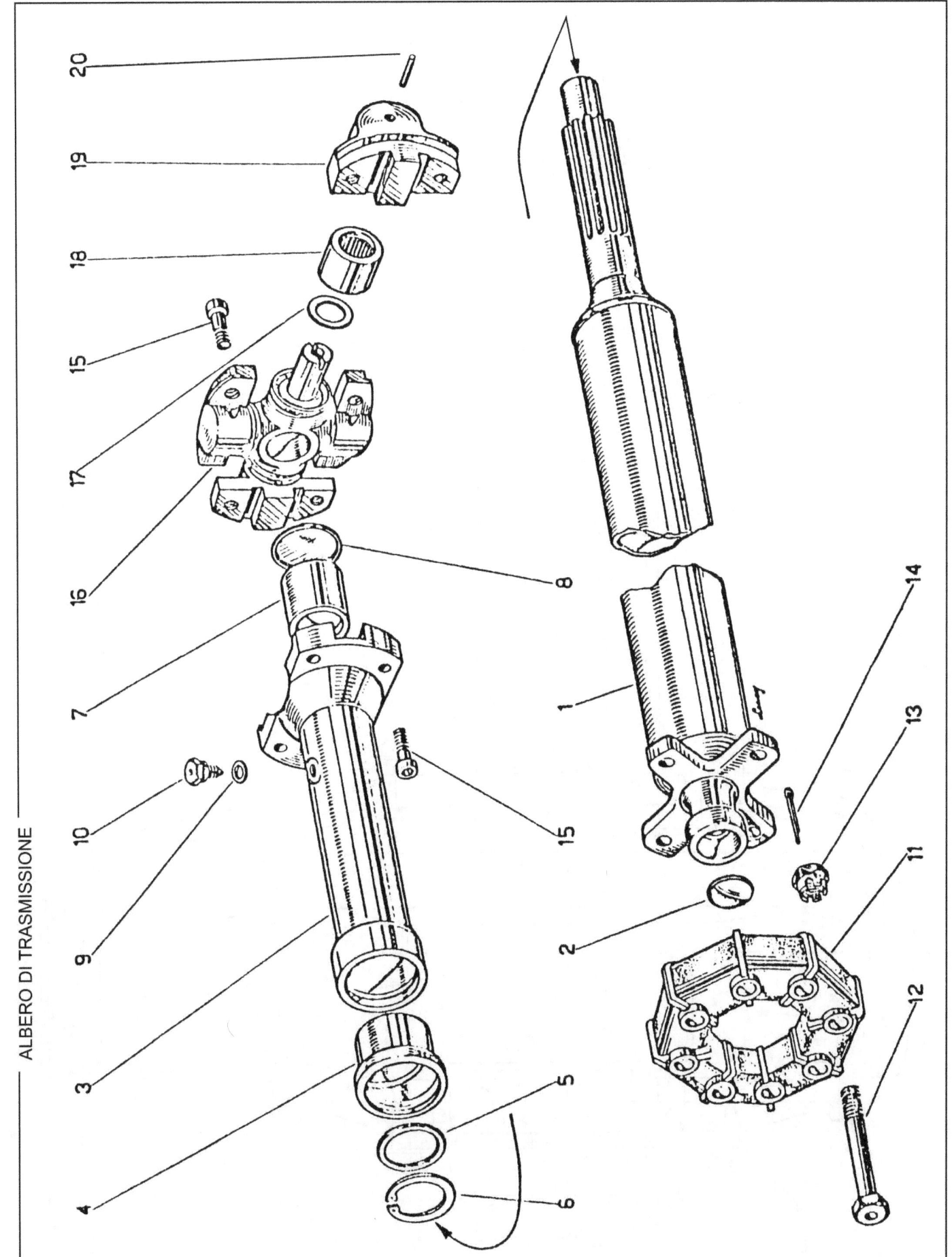

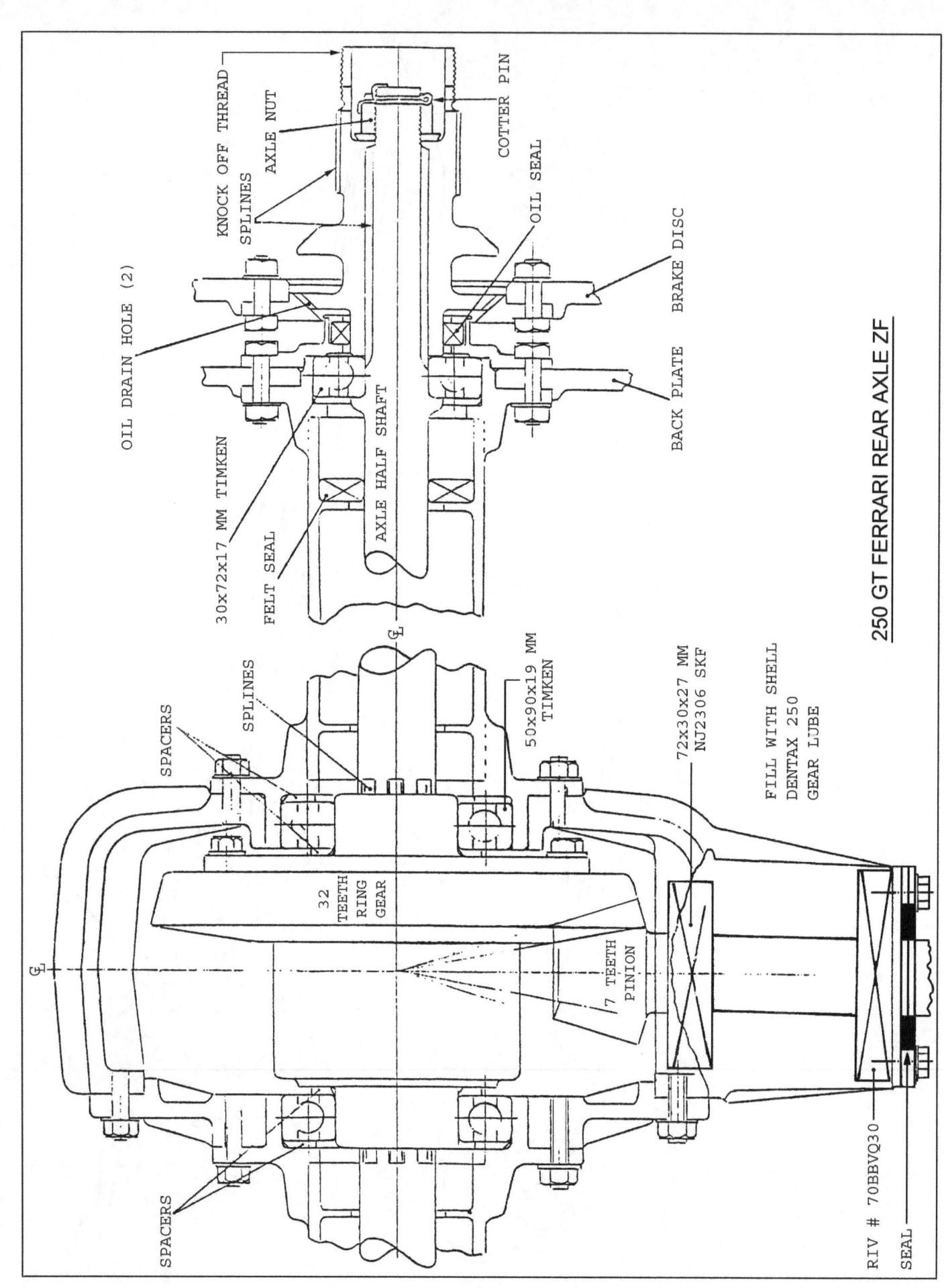

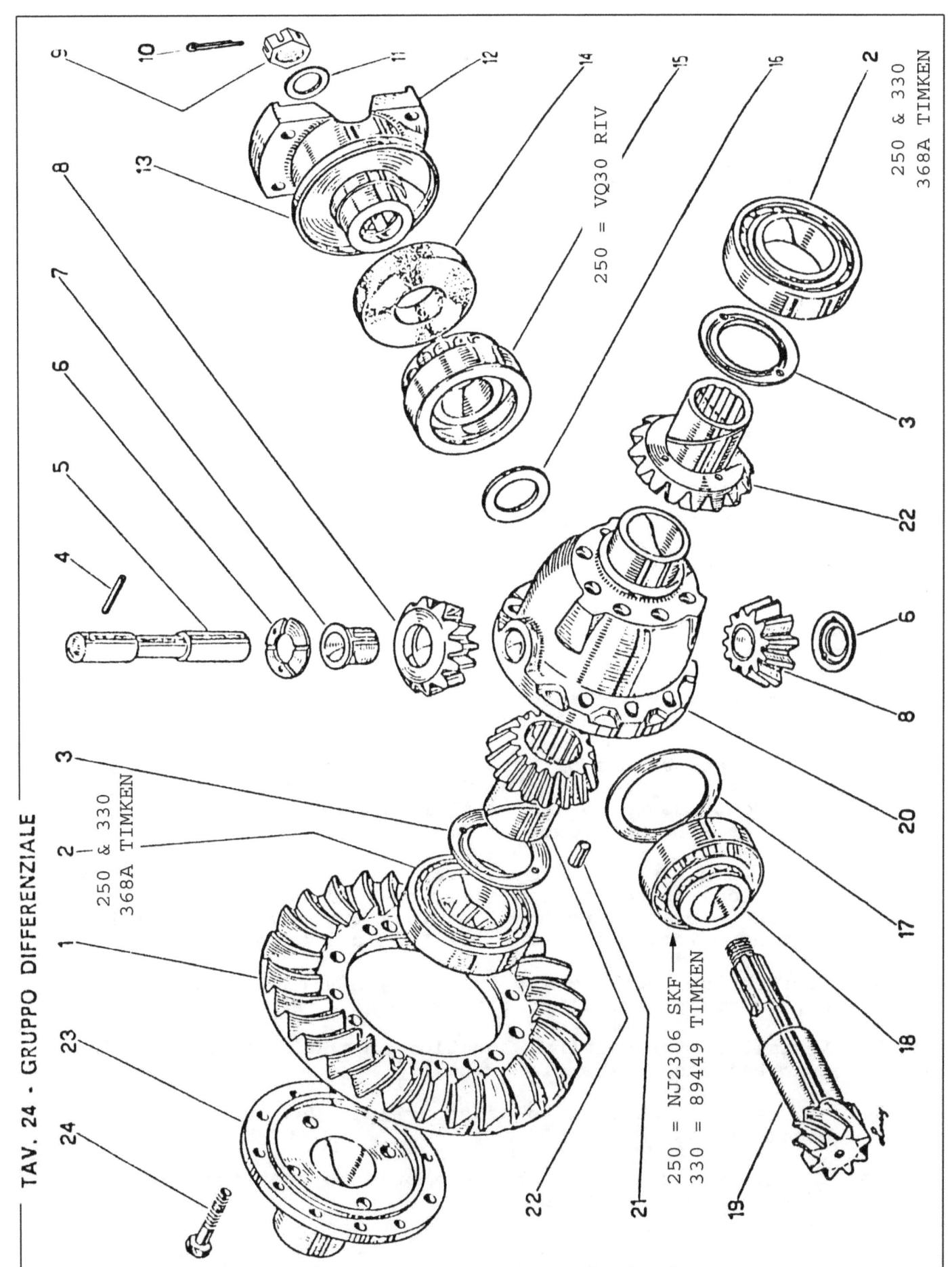

TAV. 24 - GRUPPO DIFFERENZIALE

# SHOCK ABSORBERS

## SPECIFICATIONS

Manufacture:       Koni

Front:             231 lbs. Extension, 44 lbs. Compression

Rear:              176 lbs. Extension, 44 lbs. Compression

Fluid Capacity:    1/2 Pint

Series:            82

## MAINTENANCE:

Every 1,000 Miles  -  Check for proper operation

Every 3,000 Miles  -  Check mounting bolts for tightness

Every 5,000 Miles  -  Check fluid

## ADJUSTMENT

1. After removing the shock absorber from the car, hold it in an upright position and slowly compress it by hand.

2. Holding the bottom of the shock absorber firmly, slowly rotate the top counterclockwise, continuing to compress it. Continue to rotate until no further motion is possible. Do not force or damage will occur.

3. Shock is now at minimum setting (soft ride). By rotating in a clockwise direction, damping will be heavier (firm). Do not increase more than one full turn over the previous setting. Total adjustment is two full turns.

4. Stretch the shock to its full length without turning; any rotation will alter the settings.

5. Install in the original location on the vehicle. Front and rear shocks are different.

6. Tighten all mounting bolts securely. New lock nuts are recommended.

NOTE:   All shocks on the vehicle should be adjusted the same number of turns.

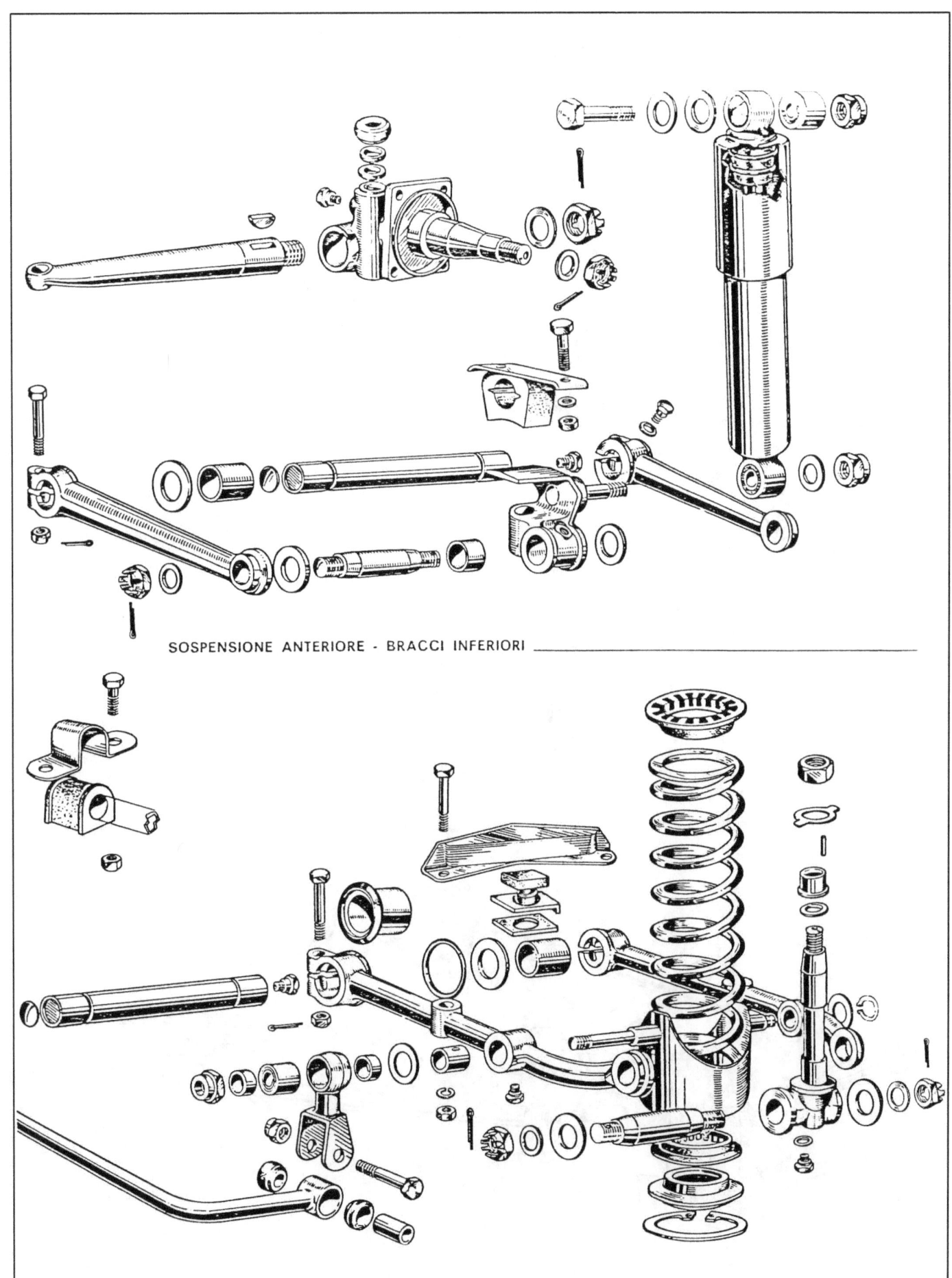

SOSPENSIONE ANTERIORE - BRACCI INFERIORI

# STEERING BOX AND STEERING LINKAGE

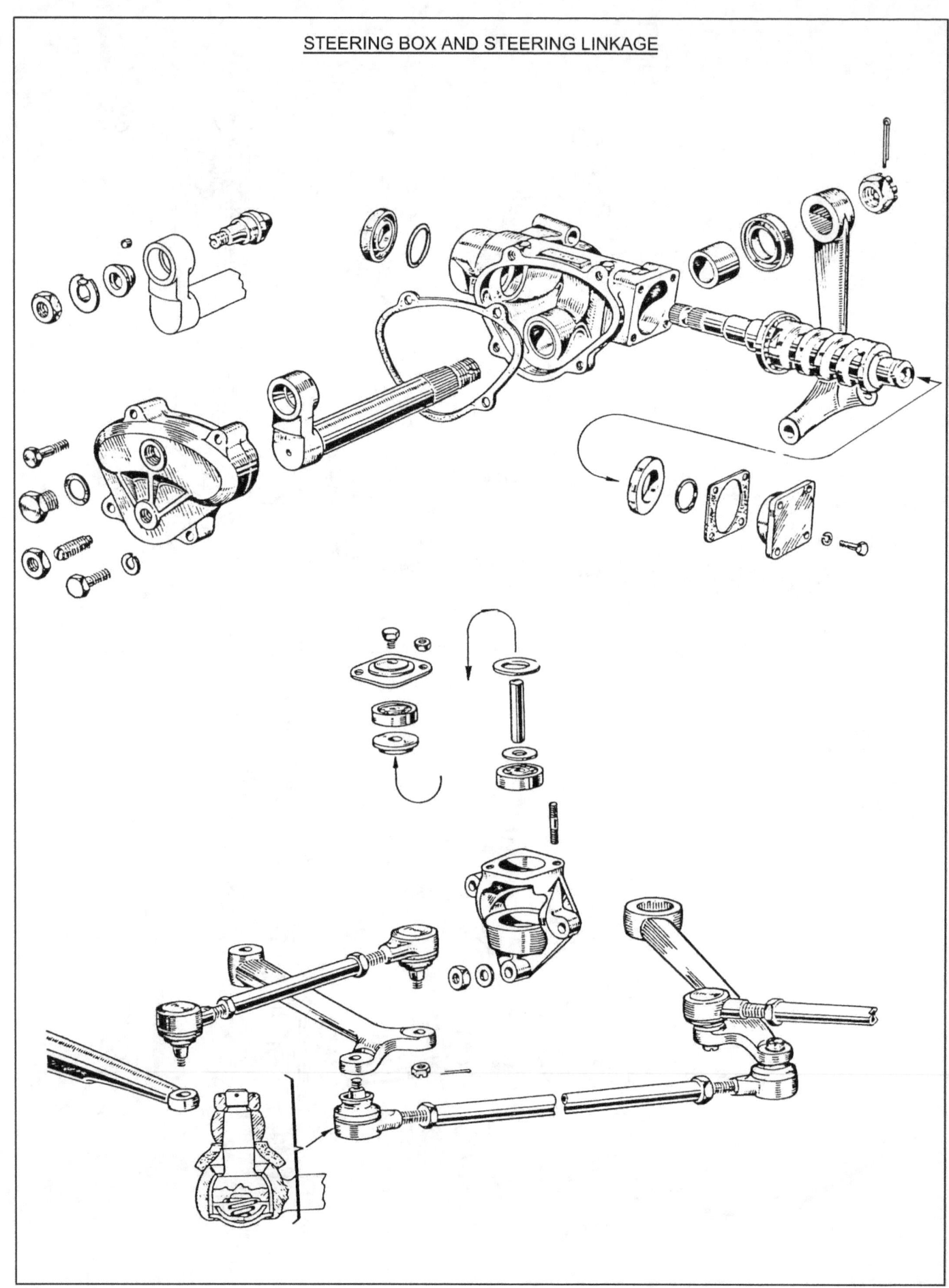

## 250GT ~ ALIGNMENT, WHEELS AND TIRES

| | |
|---|---|
| FRONT WHEEL ALIGNMENT | 124 - 125 |
| TOE IN, CAMBER AND CASTER - FACTORY SPECIFICATIONS | 126 - 127 |
| TIRES | 128 |
| WHEELS | 129 |

## FRONT WHEEL ALIGNMENT

Wheel alignment must be carried out by a specialist having equipment capable of performing the tests indicated.

### CAMBER

The camber with a static load is 1°; this is not adjustable.

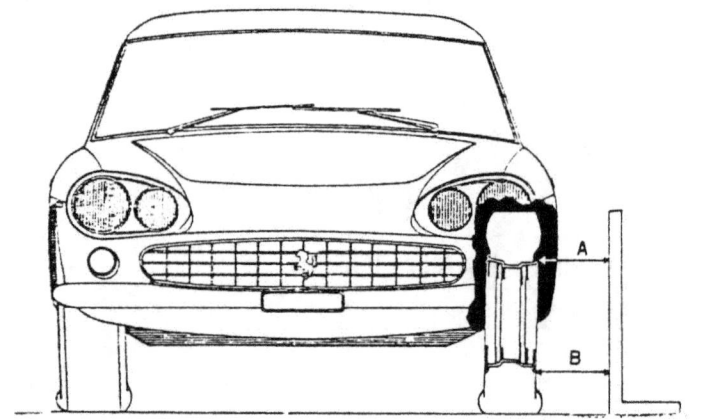

### CHECK MEASUREMENTS

B = A + 1/4" Min.
B = A + 3/8" Max.

(B = A + 6mm Min.)
(B = A + 9mm Max.)

### TOE-OUT

To adjust the toe-out, place the wheels in a straight ahead position by aligning the reference marks on the steering box and steering column. Note, the lower steering wheel spoke should be in the vertical position. Hold the steering wheel in this position. Proceed as follows:

1. Screw the steering side track rod in or out, setting the corresponding wheels in the straight ahead position, or 0" toe-out.

2. Measure the length of that steering side track-rod and adjust the opposite track-rod to the same length.

3. Then, by screwing the tie rod, in or out, bring the offside wheel into a straight ahead position.

# FRONT WHEEL ALIGNMENT (continued)

4. Lengthen both track-rods equally so as to obtain the required toe-out.

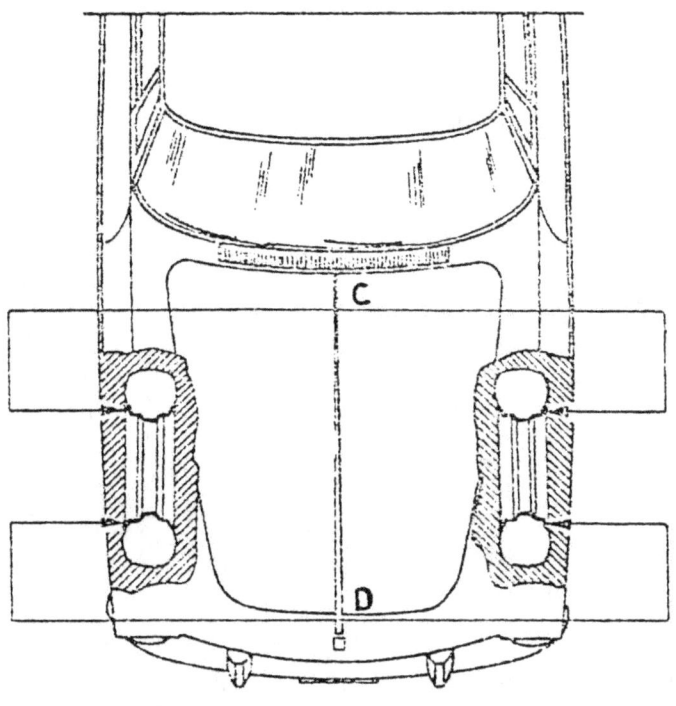

TOE-OUT MEASUREMENTS

250GT - D = Ct.08"

330GT - D = Ct.06"

## TRACK AND TIE-ROD

As measured from the center of the ball joint, the track rod length is 10.354" ± .08. If these dimensions are not obtainable, the front end may have been damaged.

View of the steering gear.
1 - Steering box; 2 - Idler arm and bracket; 3 - Tie-rod; 4 - Track rods; 5 - Turning circle stops.

# Ferrari — S.A.T.

**TECHNICAL INFORMATION**

N° 109/5
DATA: 8-11-73  F.IRMA: ─

**TABLE OF TOE-IN, CAMBER AND CASTER VALUES FOR WHEELS**

P.S. ALL DIMENSIONS ARE TO BE MADE AT STATIC LOAD (WITH TWO PASSENGERS, FULL TANK, SPARE WHEEL AND KIT)
VALUES MUST BE WITHIN TOLERANCES BUT THE SAME FOR RIGHT AND LEFT SIDE.

| CAR TYPE | FIGURE-1 FRONT WHEELS CAMBER $\alpha$ | FIGURE-2 REAR WHEELS CAMBER $\beta$ | FIGURES 3 AND 4 TOE-IN OR TOE-OUT | FIG 5 KING-PIN & CASTER |
|---|---|---|---|---|
| BERLINETTA LUSSO 250 | B=A+5,54 mm (Min), $\alpha$=+50' min <br> B=A+8,5 mm (Max), $\alpha$=+1°15' Max | C=D (Min) <br> C=D (Max) | E=F (Min); E=F-2÷3 mm (Max) <br> G=H (Min); G=H (Max) | 2°30' | 9° |
| CALIFORNIA 250 GTE/2+2 330 GT/2+2 | B=A+5,54 mm (Min), $\alpha$=+50' min <br> B=A+8,62 mm (Max), $\alpha$=+1°25' max | C=D (Min) <br> C=D (Max) | E=F (Min); E=F-2 mm (Max) <br> G=H | 2°30' | 9° |
| 275 GTB/5 275 GTB/4 | A=B+0 mm (Min); $\alpha$=0° min <br> A=B-2,02 mm (Max); $\alpha$=-0°20' max | C=D+9,97 mm (Min), $\beta$=-0°50' min <br> C=D+1,76 mm (Max), $\beta$=-1°15' max | E=F-4 mm (Min); E=F-5 mm (Max) <br> G=H-4 mm (Min); G=H-6 mm (Max) | 2°30' | 9° |
| 330 GTC/S | A=B-0 mm (Min); $\alpha$=0° min <br> A=B-2,06 mm (Max); $\alpha$=+0°20' max | C=D-5,17 mm (Min), $\beta$=-0°50' min <br> C=D-7,76 mm (Max), $\beta$=-1°15' max | E=F-6 mm (Min); E=F-5 mm (Max) <br> G=H-5 mm | 2°18'15" | 9° |
| 365 GT/2+2 | B=A-3,32 mm (Min), $\alpha$=+0°30' min <br> B=A-6,64 mm (Max); $\alpha$=+1° max | C=D-7,14 mm (Min), $\beta$=-1°5' min <br> C=D-10,5 mm (Max), $\beta$=-1°35' max | E=F (Min); E=F-3 mm (Max) <br> G=H | 2°30' | 9° |
| DINO 206 GT DINO 246 GT | A=B-1,55 mm (Min), $\alpha$=+0°05' min <br> A=B-3,10 mm (Max), $\alpha$=+0°30' max | C=D-7,75 mm (Min), $\beta$=-1°15' min <br> C=D-9,31 mm (Max), $\beta$=-1°30' max | E=F (Min); E=F-3 mm (Max) <br> G=H (Min); G=H-3 mm (Max) | 3°0'÷4°00' | 9° |
| 365 GTC/S | A=B-0 mm (Min), $\alpha$=0° min <br> A=B-2,08 mm (Max); $\alpha$=+0°20' max | C=D+10,34 mm (Min), $\beta$=-1°30' min <br> C=D (...) mm (Max), $\beta$=-2° max | E=F-4 mm (Min); E=F-5 mm (Max) <br> G=H-5 mm | 2°18'15" | 9° |

P.S. ALL DIMENSIONS ARE TO BE MADE AT STATIC LOAD (WITH TWO PASSENGERS, FULL TANK, SPARE WHEEL AND KIT). VALUES MUST BE WITHIN TOLERANCES BUT THE SAME FOR RIGHT AND LEFT SIDE.

| CAR TYPE | FIGURE - 1 FRONT WHEELS CAMBER $\alpha$ | FIGURE - 2 REAR WHEELS CAMBER $\beta$ | FIGURES - 3 AND 4 TOE-IN OR TOE-OUT | FIG 5 KING-PIN & CASTER |
|---|---|---|---|---|
| 365 GTB/4 | A=B-5.5mm (Min); $\alpha$ = +0°30' min<br>A=B-7.5mm (Max); $\alpha$ = +1°10' max | C=D-15mm (Min); $\beta$ = -2°45' min<br>C=D-16.5mm (Max); $\beta$ = -2°30' max | E=F-2mm (Min); E=F-3mm (Max)<br>G=H-2mm (Min); G=H-3mm (Max) | 1°30'  9° |
| 365 GTC/4 | A=B-5mm (Min); $\alpha$ = +0°40' min<br>A=B-7mm (Max); $\alpha$ = +1° max | C=D-10mm (Min); $\beta$ = -1°20' min<br>C=D-12mm (Max); $\beta$ = -1°40' max | E=F-2mm (Min); E=F-3mm (Max)<br>G=H | 3°  9° |
| 365 GT4/2+2 | A=B-5mm (Min); $\alpha$ = +0°40' min<br>A=B-7mm (Max); $\alpha$ = +1° max | C=D-10mm (Min); $\beta$ = -1°20' min<br>C=D-12mm (Max); $\beta$ = -1°40' max | E=F-2mm (Min); E=F-3mm (Max)<br>G=H | 3°  9° |
| 365 GT4/BB | A=B-3.4mm (Min); $\alpha$ = +0°30' min<br>A=B-6mm (Max); $\alpha$ = +0°50' max | C=D-10.1mm (Min); $\beta$ = -1°30' min<br>C=D-13mm (Max); $\beta$ = -1°50' max | E=F-1mm (Min); E=F-3mm (Max)<br>G=H-2mm (Max); G=H-4mm (Max) | 4°  9° |
| DINO 308 GT4 | A=B-1mm (Min); $\alpha$ = +0°40' min<br>A=B-3.25mm (Max); $\alpha$ = +0°50' max | C=D-8.6mm (Min); $\beta$ = -1°20' min<br>C=D-15.75mm (Max); $\beta$ = -1°40' max | E=F-1mm (Min); E=F-3mm (Max)<br>G=H-2mm (Min); G=H-4mm (Max) | 4°  9°30' |

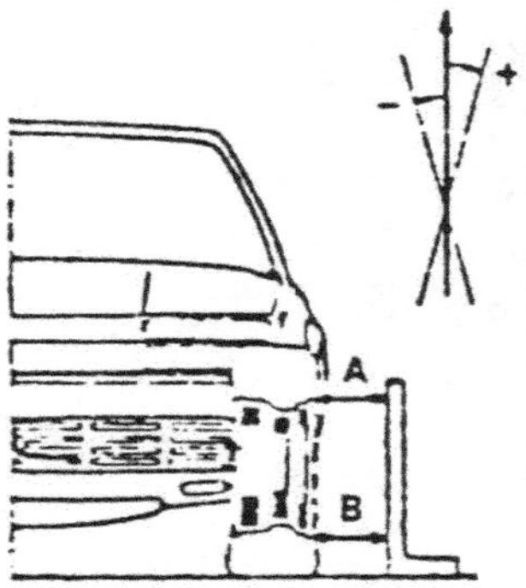

FIG. 1 - FRONT WHEELS CAMBER ($\alpha$)

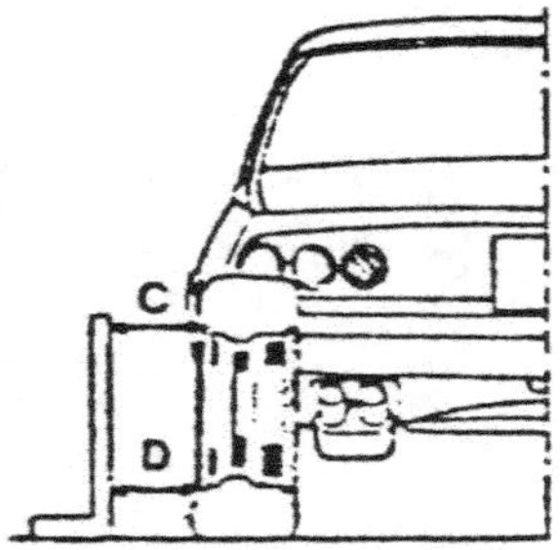

FIG. 2 - REAR WHEELS CAMBER ($\beta$)

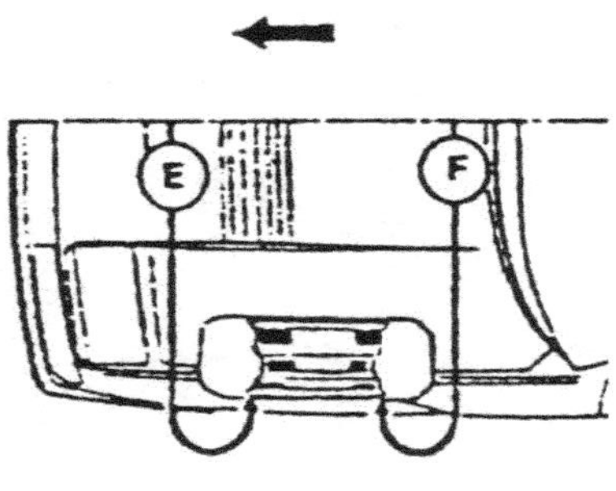

FIG. 3 - FRONT WHEELS TOE-IN

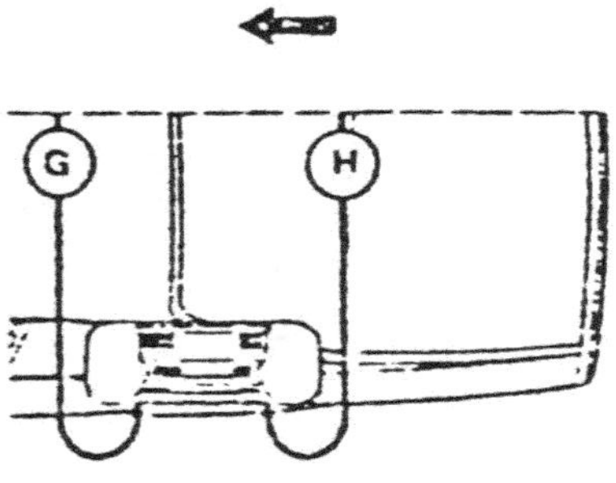

FIG. 4 - REAR WHEELS TOE-IN

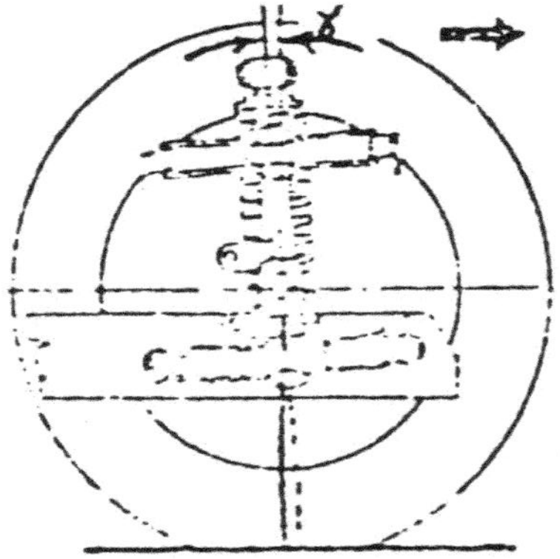

FIG. 5 - CASTER (N) FRONT

## WHEELS AND TIRES

### Ferrari Circolare Tecnica No. 108/3

| Pressures in lb/sq in | Low Speed Front | Low Speed Rear | High Speed Front | High Speed Rear |
|---|---|---|---|---|
| **250 GTE 2+2** | | | | |
|     Pirelli 185 x 15 Cinturato | 24.3 | 30.0 | 28.5 | 32.9 |
|     Dunlop 700/15 | 24.3 | 30.0 | 28.5 | 32.9 |
| **330 GT 2+2** | | | | |
|     Pirelli 210 x 15 NS | 31.3 | 34.1 | 37.1 | 40.0 |
|     Pirelli 205 HR 15 Cinturato HS | 31.3 | 34.1 | 37.1 | 40.0 |
|     Michelin 205 x 15 | 27.1 | 32.9 | 30.0 | 35.7 |
| **275 GTB** | | | | |
|     Pirelli 210 x 14 HS | 27.1 | 30.0 | 34.0 | 37.1 |
|     Dunlop 205 x 14 HR-SP | 27.1 | 32.9 | 37.1 | 42.6 |
| **275 GTS** | | | | |
|     Pirelli 210 x 14 HS | 27.1 | 30.0 | 30.0 | 32.9 |
|     Dunlop 205 x 14 HR-SP | 29.0 | 32.9 | 38.0 | 42.6 |
| **275 GTB/4** | | | | |
|     Pirelli 210 x 14 HS | 27.1 | 30.0 | 34.0 | 37.1 |
|     Michelin 205 VR 14 x or XWX | 27.1 | 32.9 | 30.0 | 37.1 |
|     Dunlop 205 x 14 HR-SP | 27.1 | 32.9 | 37.1 | 42.6 |
| **330 GTC/S** | | | | |
|     Pirelli 210 x 14 HS | 27.1 | 30.0 | 30.0 | 32.9 |
|     Michelin 205 VR 14 X or XWX | 27.1 | 32.9 | 30.0 | 37.1 |
|     Dunlop 205 x 14 HR-SP | 28.5 | 32.9 | 38.0 | 42.6 |
|     Firestone 205 VR 14 Cavallino Tubeless | 27.1 | 31.3 | 31.3 | 37.1 |
| **365 GT 2+2** | | | | |
|     Michelin 215/70 VR 15 X or XWX | 27.1 | 32.9 | 34.0 | 40.0 |
|     Firestone 205 x 15 Cavallino Tubeless | 27.1 | 32.9 | 30.0 | 36.0 |
| **365 GTC/S** | | | | |
|     Michelin 205 VR 14 X or XWX | 27.1 | 32.9 | 30.0 | 37.1 |
| **365 GTB/4** | | | | |
|     Michelin 215/70 VR 15 X or XWX | 34.0 | 38.0 | 40.0 | 44.1 |
| **365 GTC/4** | | | | |
|     Michelin 215/70 VR 15 X or XWX | 34.0 | 38.0 | 44.1 | 44.1 |
| **365 GT4/2+2** | | | | |
|     Michelin 215/70 VR 15 X or XWX | 40.0 | 47.0 | 45.5 | 49.8 |
| **DINO 206 GT** | | | | |
|     Michelin 185 VR 14 X or XWX | 27.1 | 31.3 | 27.1 | 31.3 |
|     Michelin 205/70 VR 14 X or XWX | 27.1 | 31.3 | 27.1 | 31.3 |

**RUDGE RECORD**
con cerchio di lega leggera, per ricambio e sostituzione

**RUDGE RECORD**
**wire wheels, with light alloy rim, for replacement and substitution**

**Roues fil RUDGE RECORD,**
avec jante en alliage léger, pour rechange et substitution

**RUDGE RECORD Drahtspeichenräder,**
mit Leichtmetallfelge, als Ersatz oder zur Auswechslung

| Marca e modello della vettura<br>Marque et modéle de la voiture<br>Make and model of car<br>Marke und Modell des Wagens | Misura cerchio<br>Dimension de la jante<br>Rim size<br>Felgengrösse | Tipo mozzo centrale<br>Type moyeu central<br>Type of center hub<br>Type der Radnabe | Disegno n.<br>Plan n.<br>Drawing n.<br>Zeichnung n. | Codice<br>Code<br>Code<br>Kennzahl |
|---|---|---|---|---|
| Austin Hearly 3000 | 15 x 4 ½ | 42 | RW. 3554/Bis | 35007 |
| » Hearly 3000 | 15 x 5 | 42 | RW. 3973 | 35039 |
| | | | | |
| Ferrari 250/GT | 15 x 5 | 42 | RW. 2924 | 35068 |
| » 250/GT 2+2 | 15 x 5 ½ | 42 | RW. 3591 | 35077 |
| » 250/GT 2+2 | 15 x 6 | 42 | RW. 3690 | 35013 |
| » 250/GTO | 15 x 6 | 42 | RW. 3711 | 35015 |
| » 250/GTO | 15 x 6 ½ | 42 | RW. 3715 | 35016 |
| » 250/LM | 15 x 6 | 32 | RW. 3770 | 35069 |
| » 250/GT 2+2 | 15 x 6 ½ | 42 | RW. 3801 | 35024 |
| » 250/LM | 15 x 7 ½ | 32 | RW. 3807 | 35070 |
| » 250/GTO | 15 x 7 ½ | 42 | RW. 3808 | 35071 |
| » 330/GT | 15 x 7 | 42 | RW. 3812 | 35026 |
| » 275/GTS | 14 x 6 ½ | 32 | RW. 3874 | 35034 |
| » 275/GTS | 15 x 6 ½ | 32 | RW. 3918 | 35072 |
| 275/GTS-330 | 14 x 7 | 32 | RW. 4039 | 35050 |
| Ferrari<br>250/GT-365/GT 2+2 | 15 x 7 ½ | 42 | RW. 4075 | 35053 |
| | | | | |
| Jaguar MK2 - Tipo E | 15 x 5 | 52 | RW. 3585 | 35009 |
| » MK2 - Tipo E | 15 x 7 | 52 | RW. 3886 | 35036 |
| » MK2 - Tipo E | 15 x 6 | 52 | RW. 3991 | 35041 |
| » MK2 - Tipo E | 15 x 6 ½ | 52 | RW. 4012 | 35045 |
| | | | | |
| Lamborghini 350/GT | 15 x 6 | 42 | RW. 3831 | 35029 |
| » 350/GT | 15 x 7 | 42 | RW. 3893 | 35073 |
| | | | | |
| Maserati 3700/GT | 16 x 6 | 42 | RW. 3823 | 35074 |
| » 3700/GT | 16 x 6 | 42 | RW. 3872 | 35075 |
| » 3700/GT | 15 x 6 ½ | 42 | RW. 3875 | 35076 |
| » 3700/GT | 15 x 6 ½ | 42 | RW. 3994 | 35042 |
| Maserati Mexico/Mistral | 15 x 6 ½ | 52/62 | RW. 4121 | 35059 |
| | | | | |
| MG/A | 15 x 4 ½ | 42 | RW. 3554/Bis | 35007 |
| MG/A | 15 x 5 | 42 | RW. 3973 | 35039 |

NOTES

## 250GT ~ MISCELLANEOUS

| | |
|---|---|
| COMMON TORQUE SPECIFICATIONS | 132 |
| METRIC TO USA CONVERSION FACTORS | 133 |
| MECHANICAL PROPERTIES OF BOLTS | 134 |

## 250GT TORQUE SPECIFICATIONS

Models:       250GT, 250GTE, 250GTL
Engine Type:  128, 168 (B-C-D-E-F)
Years:        1956 - 1964

ENGINE ASSEMBLY TORQUES:

| | |
|---|---|
| Cylinder head nuts for inside plug 3 bolt head | 55 lb. ft. |
| Cylinder head nuts for outside plug 4 bolt head | 60-62 lb. ft. |
| Main bearing nuts (large*) | 45 lb. ft. |
| Connecting rods | 35 lb. ft. |
| Rocker Assembly/Cam Bearing Nuts | 20 lb. ft. |
| Flywheel Bolts | 35 lb. ft. |
| Pressure Plate Bolts | 30 lb. ft. |
| Spark Plugs | 22 lb. ft. |
| Bellhousing to transmission nuts | 18 lb. ft. |
| Universal Joints | 18 lb. ft. |
| Cam Cover Acorn Nuts | 5 lb. ft. |
| 8 mm Nuts on Engine Case | 13-15 lb. ft. |
| 6 mm Oil Pan Nuts | 5 lb. ft. |
| *Small main bearing nuts (on 4 bolt style) | 25 lb. ft. |

Note: Torques indicated are for clean lubricated threads in good condition, with air and thread temperatures between 60°F and 80°F. Flat washers must be free of knicks and burrs to insure proper torque.

## METRIC STANDARDS

Some Ferraris intended for European delivery have guages and specifications measured in the metric system. Except for the oil pressure guages (*), which are calibrated in meters of head pressure, the other systems are straight forward.

| METRIC STANDARD | INTO U.S. STANDARD | MULTIPLY METRIC BY | USE |
|---|---|---|---|
| Atmospheres (ATM) | Lbs. per sq. in. (PSI) | 14.70 | Tire Pressure |
| Centigrade (°C) | Fahrenheit (°F) | $(C \times \frac{9}{5}) + 32$ | Temperature |
| Cubic Centimeters (CC) | Cubic Inches (CI) | .06102 | Displacement |
| Kilometers (Km) | Miles (Mi) | .6214 | Distance |
| Kilograms (KG) | Pounds (Lbs.) | 2.205 | Weight |
| Kilometers per Hour (KPH) | Miles per Hour (MPH) | .6214 | Speed |
| Kilogram Meters (KGM) | Pound Feet (Lb. Ft.) | 7.233 | Torque |
| Liter (L) | Quarts (Qts.) | 1.057 | Liq. Measure |
| Meter Pressure* (Oilo) | Lbs. per sq. in. (PSI) | 1.42 | Oil Pressure |
| Millimeters (MM) | Inches (In.) | .03937 | Dimensions |

# MECHANICAL PROPERTIES OF BOLTS

## MAX. TIGHTENING TORQUES IN POUND FEET - REGULAR PITCH (NC)

| Thread Size → | 4 mm | 5 mm | 6 mm | 7 mm | 8 mm | 10 mm | 12 mm | 14 mm | 16 mm | 18 mm | 20 mm | 22 mm | 24 mm | 27 mm | 30 mm |
|---|---|---|---|---|---|---|---|---|---|---|---|---|---|---|---|
| Grade 2 | 1.1 | 2.3 | 3.9 | 6.5 | 10 | 20 | 34 | 54 | 80 | 114 | 162 | 202 | 245 | 360 | 500 |
| Grade 3 | 1.7 | 3.5 | 5.8 | 9.4 | 14 | 29 | 50 | 79 | 122 | 170 | 220 | 318 | 410 | 606 | 815 |
| Grade 5 | 2 | 4 | 7 | 11 | 18 | 32 | 58 | 94 | 144 | 190 | 260 | 368 | 470 | 707 | 967 |
| Grade 8 | 2.9 | 6 | 10 | 16 | 25 | 47 | 83 | 133 | 196 | 269 | 269 | 366 | 664 | 996 | 1357 |
| Super/Allen | 3.6 | 7 | 11 | 20 | 29 | 58 | 100 | 159 | 235 | 323 | 440 | 628 | 794 | 1205 | 1630 |

Torques Indicated are for Clean, Lubricated, Threads - Cold.

## MECHANICAL PROPERTIES

| GRADE (US) | | 1 | 2 | 3 | 5 | 8 | ALLEN | SUPER |
|---|---|---|---|---|---|---|---|---|
| Tensile Strength | p.s.i. Min. | 56,000 | 70,000 | 85,000 | 113,000 | 142,000 | 170,000 | 200,000 |
| | p.s.i. Max. | 78,000 | 100,000 | 113,000 | 142,000 | 170,000 | 200,000 | 230,000 |
| Yield Stress | p.s.i. | 45,000 | 56,000 | 76,000 | 91,000 | 128,000 | 153,000 | 180,000 |
| Hardness | Brinell HB | 110-170 | 140-215 | 170-245 | 225-300 | 280-365 | 330-425 | 390- + |
| | Rockwell HRC | - | - | - | 18-31 | 27-39 | 33-44 | 40-49 |

## GRADE MARKING (QUALITY)

| | 3.6 | 4.6 | 4.8 | 5.6 | 5.8 | 6.6 | 6.8 | 6.9 | 8.8 | 10.9 | 12.9 | 14.9 |
|---|---|---|---|---|---|---|---|---|---|---|---|---|
| New | | | | | | | | | | | | |
| Before | 4A | 4D | 4S | 5D | 5S | 6D | 6S | 6G | 8G | 10K | 12K | - |
| US Grade Approx. | | 1 | | 2 | | | 3 | | 5 | 8 | | |

## SCREW THREAD LENGTHS  ISO METRIC STANDARD

| SCREW LENGTH | | THREAD LENGTH |
|---|---|---|
| FROM | TO | MM |
| - | 125 MM | 2x Diameter plus 6 MM |
| 125 MM | 200 MM | 2x Diameter plus 12 MM |
| 200 MM | - | 2x Diameter plus 25 MM |

## 250GT ~ PAINT CODES AND SUPPLIERS

DITZLER          136

GLIDDEN          137

# Ditzler IMPORTED CAR COLORS

## FERRARI

Ditzler Automotive Finishes
PPG INDUSTRIES

MADE IN ITALY

ORIGINAL FINISH – ENAMEL & ACRYLIC LACQUER

| YEAR | PAINT CODE | COLOR NAME | GENERAL DESCRIPTION | CHIP NO. | DITZLER CODE | YEAR | PAINT CODE | COLOR NAME | GENERAL DESCRIPTION | CHIP NO. | DITZLER CODE |
|---|---|---|---|---|---|---|---|---|---|---|---|
| 1963-71 | 20414.A BIA | Off White (Tetratema Bianco) | Gray White – much lighter and cleaner than | 6 | 8625 | 1963-67 | 19249 36 | Medium Gray (Grigio Medio) | Medium Gray – lighter than | 62 | 32497 |
| 1971 & Prior | | Dark Ronald Black | | | 9000 | 1968-71 | 330 | LeSancy Silver Gray Poly | Medium Gray – lighter than | 67 | 32771 |
| 1963-71 | 2033.6A AZZ | Blue Poly (Hyperion Azzurro) | Medium Gray Blue – much cleaner, lighter than (with Poly) | 14 | 13093 | 1968-71 | | Mahmoud Gray Poly | Medium Gray – darker (with Poly) | 70 | 32779 |
| 1963-71 | 48 | Ribot Blue Poly | Medium Blue – lighter, bluer than (with Poly) | 3 | 13094 | 1968-71 | | Ortello Gray Poly | Medium Dark Gray – darker (with Poly) | 37 | 32780 |
| 1963-67 | 45 | Silver Blue Poly | Medium Silver Blue Gray – lighter, bluer than | 16 | 13095 | 1968-71 | | Molvedo Turquoise | Medium Dark Green – slightly lighter | 114 | 43997 |
| 1968-71 | | Gainsborough Celeste Poly | Silver Blue – slightly lighter than | 16 | 13770 | 1968-71 | | Blenheim Green Poly | Medium Dark Green – darker, richer than | 52 | 43998 |
| 1968-71 | | Tourbillon Blue Poly | Dark Blue – darker, bluer than | 47 | 13771 | 1968-71 | | Bahram Green Poly | Medium Light Green – darker, richer than | 64 | 43999 |
| 1968-71 | | Caracalla Blue Poly | Dark Blue – deeper blue | 25 | 13772 | 1968-71 | | Seabird Green | Dark Green | 65 | 44000 |
| 1968-71 | | Bright Blue Poly (Gladiateur Azzurro) | Bright Blue – much brighter and bluer than | 96 | 13773 | 1963-69 | | Bull Lea Maroon | Maroon | 75 | 50710 |
| 1963-67 | 20325.S MAR | Cordovan (Marrone) | Dark Brown – browner than | 84 | 22612 | 1968-71 | | Blandford Violet | Purple Violet | | 50814 |
| 1963-67 | 20451.S NOC | Tan (Nocciola) | Medium Tan – darker – no Poly | 113 | 22614 | 1968 | | Race Car Red | Light Red – slightly darker than | 33 | 70797 |
| 1968-71 | | Colorado Brown Poly | Bronze Gold – darker, richer, browner than | 19 | 23149 | 1963-68 | | Red | Bright Red – richer, darker | 99 | 71508 |
| 1968-71 | | Kelso Gold Poly | Dark Gold – darker than | 89 | 23150 | 1968 | | Red | Dark Red – much darker | 33 | 71727 |
| 1968-71 | | Nashrullah Gold Poly | Medium Beige – darker (with Poly) | 24 | 23151 | 1968-71 | 19374 | Rosso Red | Medium Dark Red – (redder than 71727) | | 71745 |
| 1963-67 | 20152.S GRI | Dark Gray (Grigio Scuro) | Dark Gray – darker than | 69 | 32496 | 1968-71 | | The Tetrarch Cream | Pale Cream – lighter than | 13 | 81729 |
| | | | | | | 1968-71 | | Man O'War Yellow | Pale Yellow – cleaner than | 105 | 81730 |

DITZLER AUTOMOTIVE FINISHES — PPG INDUSTRIES — DETROIT, MICHIGAN 48235

# GLIDDEN SALCHI S.P.A. PAINT

No less than 27 colors are available in acrylic enamels for contemporary Ferraris from Carrozzeria Scaglietti, produced by Glidden Salchi SPA of Milano.

| Name | Number | Description |
|---|---|---|
| Rosso Cordobo Metallizzato | 106-R-7 | Dark Red Metallic |
| Rosso Rubino | 106-R-12 | Medium Red Metallic |
| Nocciola Metallizzato | 106-M-27 | Bronze Metallic |
| Oro Chiaro Metallizzato | 106-Y-19 | Clear Gold Metallic |
| Bleu Notte Metallizzato | 106-A-31 | Light Blue-Green Metallic |
| Celeste Chiaro Metallizzato | 106-A-26 | Clear Blue Metallic |
| Celeste Metallizzato | 106-A-16 | Blue-Gray Metallic |
| Azzurro Metallizzato | 106-A-32 | Silver-Blue Metallic |
| Verde Pino Metallizzato | 106-G-30 | Pine Green Metallic |
| Verde Medio Metallizzato | 106-G-29 | Medium to Light Green Metallic |
| Grigio Ferro Metallizzato | 106-E-8 | Iron Gray Metallic |
| Grigio Notte Metallizzato | 106-E-28 | Warm Silver Gray |
| Argento Auteil | 106-E-1 | Silver |
| Marrone Ferrari | 20-M-189 | Dark Red-Brown |
| Amaranto Ferrari | 20-R-188 | Medium Maroon Red |
| Rosso Ferrari | 20-R-187 | Medium Red |
| Rosso Chiaro Ferrari | 20-R-190 | Bright Red |
| Nero | 20-B-50 | Black |
| Bleu Ultrascuro | 20-A-174 | Navy Blue |
| Bleu Ferrari | 20-A-185 | Dark Blue |
| Azzuro La Plata | 20-A-167 | Clear Light Blue |
| Verde Scuro Ferrari | 20-G-186 | Very Dark Green |
| Grigio Brighton | 20-E-166 | Dark Gray |
| Bianco | 96-W-157 | Off White or Light Gray |
| Bianco Polo Park | 20-W-152 | Pure White |
| Avorio | 20-Y-153 | Ivory |
| Giallo Fly | 20-Y-191 | Bright Yellow |

## REMOVING PAINT

For removing those stray sprays of paint from glass, rubber, gaskets, and chrome trim, 00 steel wool does a good job without scratching the surface. This extra-fine steel wool can be used to polish chrome trim and plastic parts.

## Epilog

A Ferrari is built logically on sound, time proven methods. Ferrari SPA did not resort to quick or tricky design practices. These early cars were built by hand, and the assembly line workers never used strange tools or unique tactics to construct them. Thus, in repairing and maintaining an old Ferrari, the logic and principles that apply to addressing any problem are relevant. With this thought in mind, repairs can usually be accomplished by the owner. These simple techniques can, therefore, help keep your Ferrari a highly valued automobile for many years to come.

Jim Riff ~ 2006

## VELOCEPRESS MANUALS – AUTOMOBILE BY MAKE

ALFA ROMEO GIULIA WORKSHOP MANUAL 1300 TO 2000cc 1962-1975
ALFA ROMEO GIULIA TECH MANUAL CARBURETED CARS FROM 1962
ALFA ROMEO GIULIA TECH MANUAL FUEL INJECTED CARS FROM 1969
ALFA ROMEO GIULIETTA & GIULIA 750 & 101 SERIES 1955-1965 WSM
AUSTIN-HEALEY SPRITE & MG MIDGET WORKSHOP MANUAL 1958-1971
BMW 600 LIMOUSINE FACTORY WORKSHOP MANUAL
BMW 600 LIMOUSINE OWNERS HAND BOOK & SERVICE MANUAL
BMW 2000 & 2002 1966-1976 WORKSHOP MANUAL
CORVAIR 1960-1969 WORKSHOP MANUAL
CORVETTE V8 1955-1962 WORKSHOP MANUAL
FERRARI HANDBOOK ROAD & RACE CARS (SERVICE/SPECS) 1948-1958
FERRARI 250/GT SERVICE & MAINTENANCE MANUAL 1956-1965
FIAT 500 FACTORY WORKSHOP MANUAL 1957-1973
FIAT 600, 600D & MULTIPLA FACTORY WORKSHOP MANUAL 1955-1969
JAGUAR E-TYPE 3.8 & 4.2 SERIES 1 & 2 WORKSHOP MANUAL
JAGUAR MK 7, 8, 9 & XK120, 140, 150 WORKSHOP MANUAL 1948-1961
METROPOLITAN FACTORY WORKSHOP MANUAL
MGA & MGB OWNERS HANDBOOK & WORKSHOP MANUAL
MG MIDGET TC, TD, TF & TF1500 WORKSHOP MANUAL
PORSCHE 356 1948-1965 WORKSHOP MANUAL
PORSCHE 911 2.0, 2.2, 2.4 LITRE 1964-1973 WORKSHOP MANUAL
PORSCHE 911 2.7, 3.0, 3.2 LITRE 1973-1989 WORKSHOP MANUAL
PORSCHE 912 WORKSHOP MANUAL
PORSCHE 914/4 & 914/6 1.7, 1.8, 2.0 LITRE 1970-1976 WSM
TRIUMPH TR2, TR3, TR4 1953-1965 WORKSHOP MANUAL
VOLKSWAGEN TRANSPORTER, TRUCKS & WAGONS 1950-1979 WSM
VOLVO 1944-1968 ALL MODELS WORKSHOP MANUAL

## VELOCEPRESS TECHNICAL BOOKS - AUTOMOBILE

HOW TO BUILD A FIBERGLASS CAR
HOW TO BUILD A RACING CAR
HOW TO RESTORE THE MODEL 'A' FORD
MASERATI OWNER'S HANDBOOK
PERFORMANCE TUNING THE SUNBEAM TIGER
SOUPING THE VOLKSWAGEN
SOLEX CARBURETORS (EMPHASIS ON UK & EU AUTOMOBILES)
SU CARBURETORS (EMPHASIS ON UK AUTOMOBILES)
WEBER CARBURETORS (EMPHASIS ON ALFA & FIAT)

## VELOCEPRESS BOOKS & GUIDES - AUTOMOBILE

COMPLETE CATALOG OF JAPANESE MOTOR VEHICLES
FERRARI 308 SERIES BUYER'S AND OWNER'S GUIDE
FERRARI BROCHURES AND SALES LITERATURE 1968-1989
FERRARI SERIAL NUMBERS PART I - ODD NUMBERS TO 21399
FERRARI SERIAL NUMBERS PART II - EVEN NUMBERS TO 1050
HENRY'S FABULOUS MODEL "A" FORD
MASERATI BROCHURES AND SALES LITERATURE

## VELOCEPRESS BOOKS – AUTO RACING

CARRERA PANAMERICANA - MEXICAN ROAD RACE (BOOK OF)
DIALED IN - THE JAN OPPERMAN STORY
VEDA ORR'S NEW REVISED HOT ROD PICTORIAL

www.VelocePress.com

## VELOCEPRESS MANUALS – MOTORCYCLE BY MAKE

AJS 1932-1948 SINGLES & TWINS 250cc THRU 1000cc (BOOK OF)
AJS 1945-1960 SINGLES 350cc & 500cc MODELS 16 & 18 (BOOK OF)
AJS 1955-1965 SINGLES 350cc & 500cc (BOOK OF)
AJS 1957-1966 FACTORY WSM - ALL SINGLES & TWINS
ARIEL UP TO 1932 (BOOK OF)
ARIEL 1932-1939 PREWAR MODELS (BOOK OF)
ARIEL 1933-1951 (WORKSHOP MANUAL)
ARIEL 1939-1960 4 STROKE SINGLES (BOOK OF)
ARIEL 1958-1964 LEADER & ARROW FACTORY WSM & PARTS LIST
ARIEL 1958-1964 LEADER & ARROW (BOOK OF)
BMW R26 R27 (1956-1967) FACTORY WORKSHOP MANUAL
BMW R50 R50S R60 R69S (1955-1969) FACTORY WORKSHOP MANUAL
BMW R50/5 R60/5 R75/5 (1969-1973) FACTORY WORKSHOP MANUAL
BRIDGESTONE 90 SERIES FACTORY WSM & PARTS CATALOGUE
BRIDGESTONE 175 SERIES FACTORY WSM & PARTS CATALOGUE
BRIDGESTONE 350 SERIES FACTORY WSM & PARTS CATALOGUES
BSA SERVICE SHEETS MASTER CATALOGUE ALL MODELS 1945-1967
BSA BANTAM D1 TO D7 1948-1966 FACTORY SERVICE SHEETS MANUAL
BSA BANTAM ALL MODELS FROM 1948 ONWARDS (BOOK OF)
BSA BANTAM D14 FACTORY SERVICE MANUAL
BSA DANDY FACTORY WORKSHOP MANUAL (COMPILATION)
BSA SINGLES & V-TWINS UP TO 1926 inc. 1927 SUPPLEMENT (BOOK OF)
BSA SINGLES & V-TWINS UP TO 1930 (BOOK OF)
BSA SINGLES & V-TWINS UP TO 1935 (BOOK OF)
BSA SINGLES & V-TWINS 1936-1939 (BOOK OF)
BSA C10, C11 & C12 1945-1958 FACTORY SERVICE SHEETS MANUAL
BSA OHV & SV SINGLES 250-600cc 1945-1959 (BOOK OF)
BSA C15 & B40 1958-1967 FACTORY SERVICE SHEETS MANUAL
BSA OHV & SV SINGLES 250cc (ONLY) 1954-1970 (BOOK OF)
BSA B31, B32, B33 & B34 1945-60 FACTORY SERVICE SHEETS MANUAL
BSA OHV SINGLES 350 & 500cc 1955-1967 (BOOK OF)
BSA M20, M21 & M33 1945-1963 FACTORY SERVICE SHEETS MANUAL
BSA TWINS A7 & A10 1948-1962 FACTORY SERVICE SHEETS MANUAL
BSA TWINS A7 & A10 1948-1962 (BOOK OF)
BSA TWINS A50 & A65 1962-1965 FACTORY WORKSHOP MANUAL
BSA TWINS A50 & A65 1962-1969 (SECOND BOOK OF)
DOUGLAS 1929-1939 PREWAR ALL MODELS (BOOK OF)
DOUGLAS 1948-1957 POSTWAR ALL MODELS FACTORY SHOP MANUAL
DUCATI 160cc, 250cc & 350cc OHC MODELS FACTORY SHOP MANUAL
HONDA 50cc ALL MODELS UP TO 1970 INC MONKEY & TRAIL (BOOK OF)
HONDA 90cc ALL MODELS UP TO 1966 (BOOK OF)
HONDA TWINS ALL MODELS 50cc THRU 305cc 1960-1966 (BOOK OF)
HONDA TWINS ALL MODELS 125cc THRU 450cc UP TO 1968 (BOOK OF)
HONDA C100 50cc SUPER CUB O.H.C. 1959-1962 FACTORY WSM
HONDA C110 50cc SPORT CUB O.H.C. 1960-1962 FACTORY WSM
HONDA 50-65-70-90cc O.H.C. SINGLES 1959-1983 WSM
HONDA 100-125cc SINGLES CB/CD/CL/SL/TL 1970-1984 FACTORY WSM
HONDA 125-150cc TWINS C/CS/CB/CA 1959-1966 FACTORY WSM
HONDA 125-160-175-200cc TWINS 1965-1978 WORKSHOP MANUAL
HONDA 250-305cc TWINS C/CS/CB 1961-1968 FACTORY WSM
HOHDA 250-350cc TWINS CB/CL/SL 1968-1973 FACTORY WSM
HONDA 250-360cc TWINS CB/CL/CJ 1974-1977 FACTORY WSM
HONDA 350F & 400F 4-CYLINDER 1972-1977 FACTORY WSM
HONDA 450cc TWINS CB/CL 1965-1974 K0 TO K7 WORKSHOP MANUAL
HONDA 500cc & 550cc 4-CYL 1971-1978 FACTORY WORKSHOP MANUAL
HONDA 750cc SHOC 4-CYL 1969-1978 K0~K8 WORKSHOP MANUAL
INDIAN PONYBIKE, BOY RACER & PAPOOSE ILL PARTS LIST & SALES LIT

### VELOCEPRESS MANUALS – SCOOTERS BY MAKE

BSA SUNBEAM SCOOTER WORKSHOP MANUAL 1959-1965
BSA SUNBEAM SCOOTER 1959-1965 (BOOK OF)
LAMBRETTA 1947-1957 ALL 125 & 150cc MODELS (BOOK OF)
LAMBRETTA 1957-1970 LI & TV MODELS (SECOND BOOK OF)
NSU PRIMA 1956-1964 ALL MODELS (BOOK OF)
TRIUMPH TIGRESS SCOOTER WORKSHOP MANUAL 1959-1965
TRIUMPH TIGRESS SCOOTER (BOOK OF)
VESPA 1951-1961 (BOOK OF)
VESPA 1955-1963 125 & 150cc & GS MODELS (SECOND BOOK OF)
VESPA 1955-1968 GS & SS (BOOK OF)
VESPA 1963-1972 90, 125 & 150cc (THIRD BOOK OF)

### VELOCEPRESS MANUALS – MOPEDS & MOTORIZED BICYCLES

CYCLEMOTOR (BOOK OF)
NSU QUICKLY 1953-1963 ALL MODELS (BOOK OF)
PUCH MAXI N & S MAINTENANCE & REPAIR (3 MANUAL COMPILATION)
RALEIGH MOPEDS 1960-1969 (BOOK OF)

J.A.P. ENGINES 1927-1952 & MOTORCYCLES 1934-1952 (BOOK OF)
MATCHLESS 1931-1939 ALL MODELS 250cc THRU 990cc (BOOK OF)
MATCHLESS 1945-1956 350 & 500cc SINGLES (BOOK OF)
MATCHLESS 1955-1966 350 & 500cc SINGLES (BOOK OF)
MATCHLESS 1957-1966 FACTORY WSM - ALL SINGLES & TWINS
NEW IMPERIAL ALL SV & OHV FROM 1935 ONWARDS (BOOK OF)
NORTON 1932-1939 PREWAR MODELS (BOOK OF)
NORTON 1932-1947 (BOOK OF)
NORTON 1938-1956 (BOOK OF)
NORTON 1945-1963 MODELS 16H, Big4, ES2, 19 & 50 WSM'S & PARTS
NORTON 1955-1963 MODELS 19, 50 & ES2 (BOOK OF)
NORTON 1948-1970 DOMINATOR TWINS FACTORY WSM'S & PARTS
NORTON 1955-1965 DOMINATOR TWINS (BOOK OF)
NORTON 1960-1970 TWIN CYLINDER FACTORY WORKSHOP MANUAL
NORTON 1970-1975 COMMANDO 850 & 750cc FACTORY WSM
NORTON 1975-1978 MK 3 COMMANDO 850 cc FACTORY WSM
PANTHER 1932-1958 LIGHTWEIGHT MODELS 250 & 350cc (BOOK OF)
PANTHER 1938-1966 HEAVYWEIGHT MODELS 600 & 650cc (BOOK OF)
PENTON-KTM-SACHS 1968-1975 100cc & 125cc WORKSHOP MANUAL
RALEIGH MOTORCYCLES 1919-1933 (BOOK OF)
ROYAL ENFIELD 1934-1946 SINGLES & V TWINS (BOOK OF)
ROYAL ENFIELD 1937-1953 SINGLES & V TWINS (BOOK OF)
ROYAL ENFIELD 1946-1962 SINGLES (BOOK OF)
ROYAL ENFIELD 1948-1963 500cc TWINS FACTORY WORKSHOP MANUAL
ROYAL ENFIELD 1952-1963 700cc TWINS FACTORY WORKSHOP MANUAL
ROYAL ENFIELD 1956-1966 250cc CRUSADER & 350cc NEW BULLET WSM
ROYAL ENFIELD 1958-1966 250cc & 350cc SINGLES (SECOND BOOK OF)
ROYAL ENFIELD 1962-1970 INTERCEPTOR WSM'S & PARTS (Compilation)
RUDGE 1933-1939 (BOOK OF)
SACHS 1968-1975 100cc & 125cc ENGINES WSM & M/CYCLE PARTS LIST
SUNBEAM 1928-1939 (BOOK OF)
SUNBEAM 1946-1957 S7 & S8 (BOOK OF)
SUZUKI 50cc & 80cc UP TO 1966 (BOOK OF)
SUZUKI T10 1963-1967 FACTORY WORKSHOP MANUAL
SUZUKI T20 & T200 1965-1969 FACTORY WORKSHOP MANUAL
SUZUKI TWINS 1962 ONWARDS 125-500cc WORKSHOP MANUAL
TRIUMPH 1935-1949 SINGLES & TWINS (BOOK OF)
TRIUMPH 1937-1961 SINGLES & OHV 250cc-600cc + TERRIER & CUB
TRIUMPH 1945-1955 PRE-UNIT 350cc, 500cc & 650cc TWINS WSM No.11
TRIUMPH 1945-1959 TWINS (BOOK OF)
TRIUMPH 1956-1969 TWINS (BOOK OF)
TRIUMPH 1956-1962 PRE-UNIT 500cc & 650cc TWINS WSM No.17
TRIUMPH 1957-1963 UNIT CONSTRUCTION 350-500cc WSM No.4
TRIUMPH 1963-1974 UNIT CONSTRUCTION 350-500cc FACTORY WSM
TRIUMPH 1963-1970 UNIT CONSTRUCTION 650cc FACTORY WSM
TRIUMPH 1968-1974 TRIDENT T150 & T150V FACTORY WSM
TRIUMPH 1971-1973 650cc OIL-IN-FRAME FACTORY WSM
TRIUMPH 1973-1978 750cc BONNEVILLE & TIGER FACTORY WSM
TRIUMPH 1979-1983 750cc T140, TR7 & TR65 FACTORY WSM
VELOCETTE 1925-1970 ALL SINGLES & TWINS (BOOK OF)
VELOCETTE 1933-1952 MOV-MAC-MSS RIGID FRAME FACTORY WSM
VELOCETTE 1954-1971 MSS-VENOM-THRUXTON-VIPER FACTORY WSM
VILLIERS ENGINE UP TO 1959 INC. 3 WHEELERS (BOOK OF)
VILLIERS ENGINE UP TO 1969 (BOOK OF)
VINCENT 1935-1955 (WORKSHOP MANUAL)
YAMAHA 1961-1967 YA5 & YA6 (WORKSHOP MANUAL & ILL PARTS LIST)
YAMAHA 1971-1972 JT1& JT2 (WORKSHOP MANUAL & ILL PARTS LIST)

### VELOCEPRESS MANUALS - THREE WHEELER'S

BOND MINICAR THREE WHEELER 1948-1967 (BOOK OF)
BMW ISETTA FACTORY WORKSHOP MANUAL
BSA THREE WHEELER (BOOK OF)
RELIANT REGAL THREE WHEELER 1952-1973 (BOOK OF)
VINTAGE MORGAN THREE WHEELER (BOOK OF)

### VELOCEPRESS TECHNICAL BOOKS – MOTORCYCLE

1930'S BRITISH MOTORCYCLE CARBS & ELEC COMPONENTS (BOOK OF)
1930'S BRITISH MOTORCYCLE ENGINES (OVERHAUL & MAINTENANCE)
1930'S BRITISH MOTORCYCLE GEARBOXES & CLUTCHES (BOOK OF)
CATALOG OF BRITISH MOTORCYCLES (1951 MODELS)
LUCAS ELECTRONICS BRITISH M/CYCLES REPAIR & PARTS (1950-1977)
MOTORCYCLE ENGINEERING (P.E. Irving)
MOTORCYCLE ROAD TESTS 1949-1953 (Motor Cycle Magazine UK)
SPEED AND HOW TO OBTAIN IT (Motor Cycle Magazine UK)
TUNING FOR SPEED (P.E. Irving)
WIPAC (COMBO) MANUAL NUMBER 3 + M/CYCLE & SCOOTER MANUAL

www.VelocePress.com

www.ingramcontent.com/pod-product-compliance
Lightning Source LLC
Chambersburg PA
CBHW082211300426
44117CB00016B/2754